# Evolution in the Dark

Horst Wilkens • Ulrike Strecker

# Evolution in the Dark

## Darwin's Loss Without Selection

Springer

Horst Wilkens
Centrum für Naturkunde—Zoologisches Museum
Universität Hamburg
Hamburg, Germany

Ulrike Strecker
Centrum für Naturkunde—Zoologisches Museum
Universität Hamburg
Hamburg, Germany

ISBN 978-3-662-54510-2 ISBN 978-3-662-54512-6 (eBook)
DOI 10.1007/978-3-662-54512-6

Library of Congress Control Number: 2017941083



Printed on acid-free paper

This Springer imprint is published by Springer Nature
The registered company is Springer-Verlag GmbH Germany
The registered company address is: Heidelberger Platz 3, 14197 Berlin, Germany

# Acknowledgements

Annelore Kröger patiently prepared the voluminous graphic work. The partly coloured drawings were produced by Monika Hänel. Arnhild Woltmann constantly provided ready assistance and Detlef Teege was responsible for keeping the fish in good condition for many years. We express our deepest thanks to all of them for their continuous assistance and engagement. The *Astyanax* project has been sponsored by the Deutsche Forschungsgemeinschaft since 1970. Above all we thank the Mexican people and authorities for their help and permits. If not differently assigned, the photos are by the authors.

# Contents

# Evolution in the Dark: Introduction

# 1

**Abstract**

Evolution is predominantly understood as a progressive process, and less attention is usually paid to those traits being reduced at the same time. Since Darwin, who was the first to expound what can be called 'Darwin's loss', the main agents of regression have been under dispute.

Because evolution is considered to progress over time, the features that are constructively improved during adaptive development have been the main focus of scientific research. Less attention has been paid to the fact that, at the same time, other traits lose their function and are subjected to reduction and regressive evolution. Two types of this, aphanisia and rudimentation, were proposed to occur (Sewertzoff 1931). Aphanisia structures are those that have lost their biological function at some stage during the species-specific individual life span, such as larval tadpoles and tunicates losing their tails or insect larvae completely reducing specific or even all body parts during metamorphosis. In contrast, rudimentation characterizes the process of reduction of features that have lost their biological function at all ontogenetic stages. Rudimentation develops either when a trait had lost its biological significance or when its function is replaced by another one (Sewertzoff 1931). Such processes exceed the individual life span and generally take thousands of years. Furthermore, rudimentation differs from aphanisia by the long-lasting existence of intermediate transitional stages, which may show high variability.

Whereas there is no doubt that selection plays an important role in constructive evolution, the underlying processes of rudimentation are still under dispute (Culver and Wilkens 2000; Rétaux and Casane 2013). Darwin was already aware of this problem when he stated "There remains, however, this difficulty: after an organ has ceased being used, and has become in consequence much reduced, how can it be still further reduced in size until the merest vestige is left; and how can it be finally quite obliterated?" In the face of the neodarwinian paradigm of selection as an agent



H. Wilkens, U. Strecker, *Evolution in the Dark*, DOI 10.1007/978-3-662-54512-6_1

exclusively applied to explain evolutionary processes in recent times, almost countless efforts have been made to explain 'Darwin's loss' and to detect selection, which is no longer assumed to have direct influence on rudimentation. This issue is most evident in cave-living species, in which, in particular, the loss of the eyes has fascinated biologists since Darwin. Cave animals have become prime subjects to study pheno- and genotypic principles, not only of regression but of evolution in general. They provide exceptional tools to study the process of evolution far beyond that of eye and pigment reduction. As a result of the close relationship to their surface sister forms, cave species enable comparative morphological, physiological, and behavioural analyses of regressive and constructive traits. Persisting interfertility often allows the analysis of the genetic bases and differences between surface and cave forms. Furthermore, in combination with paleobiogeography, evolutionary rates of constructive or regressive traits like the rudimentary eye can be calculated based on molecular and morphological analyses.

## References

Culver DC, Wilkens H (2000) Critical review of the relevant theories of the evolution of subterranean animals. In: Wilkens H, Culver DC, Humphreys WF (eds) Ecosystems of the world: subterranean ecosystems, vol 30. Elsevier, Amsterdam, pp 381–398

Rétaux S, Casane D (2013) Evolution of eye development in the darkness of caves: adaptation, drift, or both? EvoDevo 4(1):1–12

Sewertzoff AN (1931) Morphologische Gesetzmäßigkeiten der Evolution. Jena, Germany

# 2 The Role of Rudimentation in Evolution

**Abstract**

Processes of regression and rudimentation are deeply involved in the evolution of life and are as important as constructive evolution. They occur in every taxonomic group and concern morphological, behavioural, as well as physiological traits. For example, whales have reduced their hind legs and the pelvic girdle. The ratite birds have convergently abandoned the ability to fly and exhibit reduced wings and sternal carina. In addition, the delicate feather structure is broken down. In the Pacific island of Tahiti, where no insectivore bats exist, noctuid moths have lost the acoustic startle response. Even the gustatory system may selectively lose taste components (e.g. sweet in cats; bitter, sweet, and umami in penguins; or umami in the giant panda after changing their diet during evolution). However, from the view of human beings relying on sight as the dominant sense, the most bizarre and striking examples for rudimentation—often also characterized as degeneration or regression of traits—are provided by the loss of eyes and dark pigmentation in species living in the continuous absolute darkness of subterranean habitats like caves.

Impressive examples of rudimentation and regression are provided by the rudimentary pelvic girdles or hip bones of whales. Their anlagen are completely developed during early ontogeny but are reduced to a size no more than three percent of the total body length in adult specimens. The pelvic girdle of whales is no longer linked to the vertebral column, but lies deeply sunken, embedded in body tissue. The rudimentary hind legs may still be extant as minute remnants of the former thighs and shanks and are characterized by great variability (Abel 1908; Deimer 1977; Thewissen and Bajpai 2001) (Figs 2.1 and 2.2).

Striking evidence for rudimentation is also delivered by a large number of bird species. With the loss of the ability to fly, they have not only reduced their wing size—some of them even totally—but have also lost the flight musculature as well as the keel of the sternum, which the flight muscles attach to, and the delicate wing



H. Wilkens, U. Strecker, *Evolution in the Dark*, DOI 10.1007/978-3-662-54512-6_2

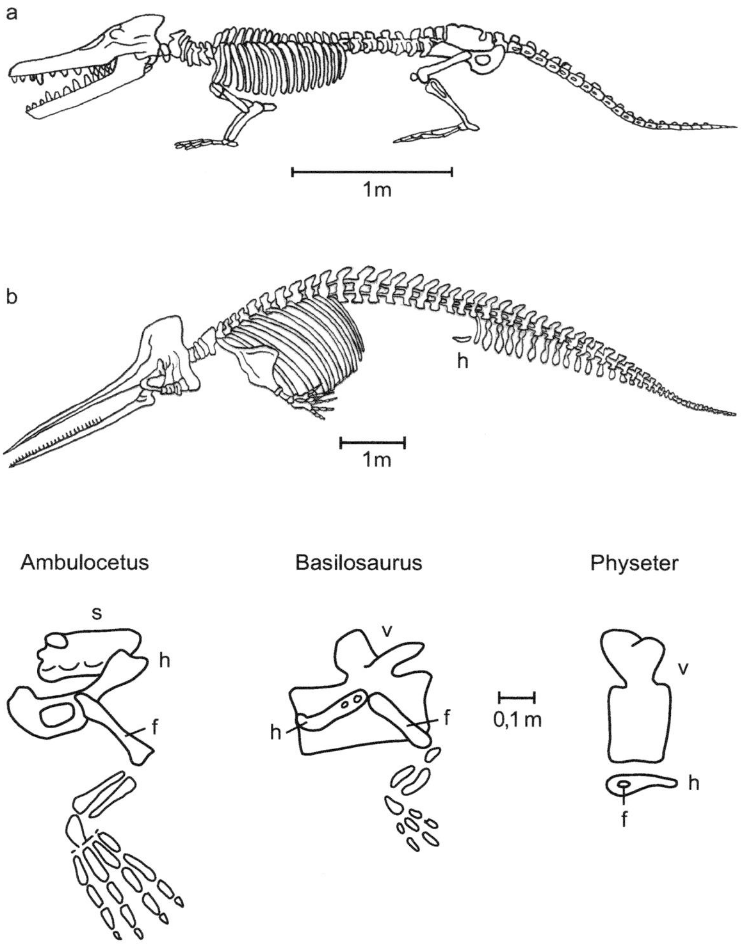


**Fig. 2.1** Regressive evolution of hip bone and hind legs in ancestral (**a**: *Ambulocetus natans*) and modern whales (**b**: Sperm Whale, *Physeter macrocephalus*). In the archaic whale *Ambulocetus*, the sacrum (*s*) consisted of four fused vertebrae and there was a strong weight-bearing joint between sacrum and hipbone (*h*). In the extinct whale *Basilosaurus* and the recent Sperm Whale (*Physeter*), the sacrum consists of a single vertebra (*v*) and there is no contact between hip bone and sacrum. In both, hip bone and femur (*f*) lie on the ventral side, far removed from the spinal column. The rudimentary hind legs are no longer weight bearing (adapted from Thewissen et al. 2001)

feather structure. The most spectacular examples are provided by the systematic group of ratite birds (Ratitae), which comprise species like the African ostrich, the Australian emus (*Dromaius novaehollandiae*; Fig. 2.3) and cassowaries, the South American nandu, the extinct elephant birds from Madagascar, and the kiwis and extinct moas from New Zealand (Fig. 2.4). Molecular genetic studies provide evidence that flightlessness in these species evolved independently and did not

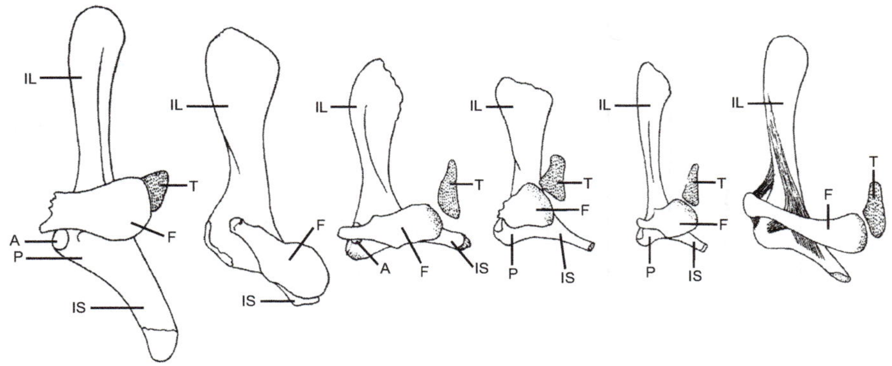


**Fig. 2.2** Variability of the left hip bone and rudimentary hind leg of six female Bowhead whales (*Balaena mysticetus*). Body length 12–15 m, length of hip bones about 0.5 m (adapted from Abel 1908). *Il* ileum, *P* pubis, *IS* ischium, *F* femur, *T* tibia

**Fig. 2.3** The ratite Australian emu (*Dromaius novaehollandiae*) exhibits reduced wings and disturbed feather structure (Photo Ulrike Strecker)

rely on common descent from a unique flightless ancestor but on convergence (Mitchell et al. 2014; Phillips et al. 2010). Flightlessness and the phenomenon of gigantism observed in most species of this group are assumed to have been facilitated by early Tertiary expansion into the diurnal herbivory niche after the extinction of the dinosaurs. Comparable evolutionary processes have also taken

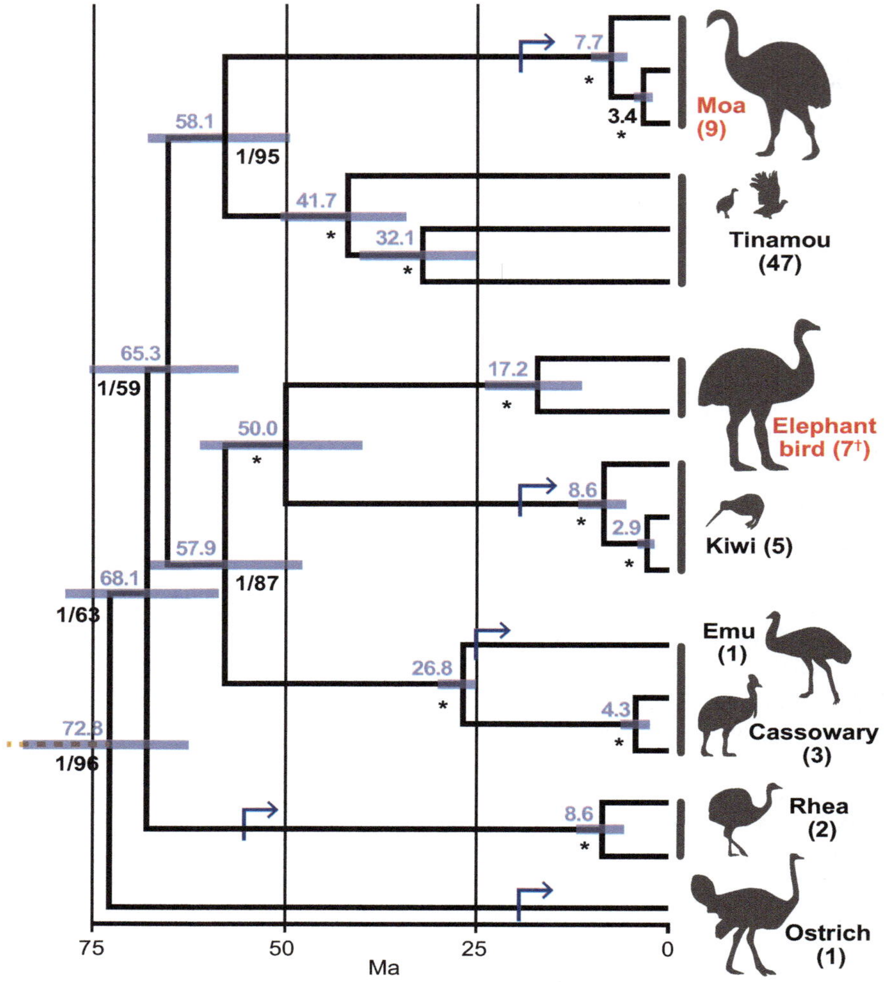


**Fig. 2.4** Phylogeny of ratite birds from mitochondrial sequence data revealed convergent origins of flightlessness. Only the tinamous species from South America are able to fly. *Numbers above branches* indicate divergence dates. *Arrows* mark the minimum date for the evolution of flightlessness in lineages for which fossil evidence is available. The scale is given in millions of years before the present. *Silhouettes* indicate the relative size of representative taxa. Species diversity for each major clade is presented in *parentheses*, with extinct groups shown in *red*. The *dagger* symbol indicates that the number of elephant bird species is uncertain (adapted from Mitchell et al. 2014)

place in other systematic avian bird families like doves (Columbiformes) and cormorants (Phalacrocoracidae) on the isolated oceanic Mascarene and Galapagos Islands, respectively. These species evolved on islands without terrestrial predators,

**Fig. 2.5** The flightless cormorant (*Phalacrocorax harrisi*) from Galapagos exhibits reduced wings and disorderly variable arrangement of primary and secondary wing feathers. This species provides an example of many birds reducing their flying abilities on isolated islands (Photo Harald Schliemann)

where the ability to escape by flight was no longer an advantage and the birds eventually became flightless (Fig. 2.5).

Fossorial mammals inhabit light-poor environments. In several species, the eyes are reduced to small size and may even be totally covered by skin (Emerling and Springer 2014; Nikitina et al. 2004). Traits like lens and retina may be subjected to structural regression. For example, in the naked mole rat (*Heterocephalus glaber*) the lens replaces the vitreous body and the retina is deformed but still consists of all sensitive layers. However, the eyes of these fossorial species cannot be completely reduced because they have still retained rudimentary biological functions like that of perceiving surface light for orientation or setting the biological clock. Analysis of the central visual system of the naked mole rat revealed that it appears to have selectively lost structures that mediate form vision while retaining those structures needed for minimal entrainment of circadian and seasonal rhythms (Crish et al. 2006).

Studying two insectivores (*Condylura cristata* and *Chrysochloris asiatica*), two rodents (*Heterocephalus glaber* and *Spalax ehrenbergi*), and a marsupial (*Notoryctes typhlops*) revealed that the fossoriality of these mammals is intimately linked to eye regression (Fig. 2.6). It was shown that a decrease in the amount of light that reaches the retina is associated with increased regression of retinal genes and that the phototransduction and visual cycle pathways degrade. The timing of retinal gene loss is associated with the entrance of mammalian lineages into

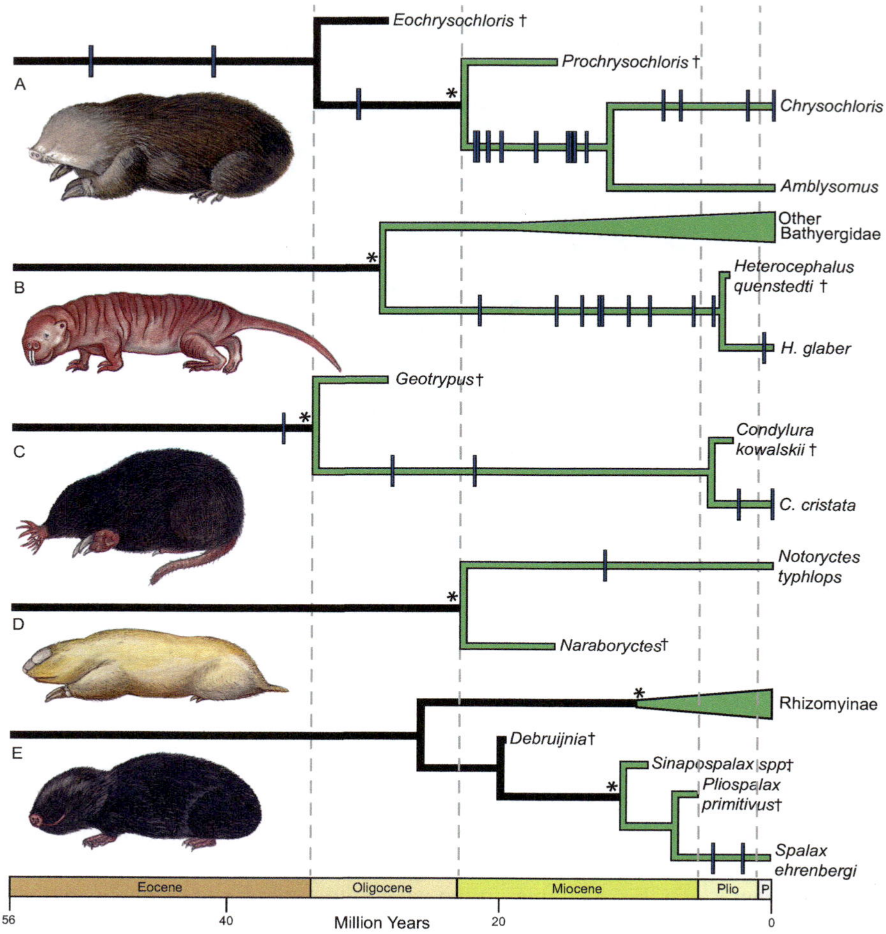


**Fig. 2.6** Pseudogenization of retinal genes in fossorial mammals: Cape golden mole (*Chrysochloris asiatica* (**a**), Chrysochloridae), Naked mole rat (*Heterocephalus glaber* (**b**), Bathyergidae), Star-nosed mole (*Condylura cristata* (**c**), Talpidae), Marsupial mole (*Notoryctes typhlops* (**d**), Notoryctidae), Blind mole rat (*Spalax ehrenbergi* (**e**), Spalacidae). *Vertical bars* represent average estimates of pseudogenization and inactivation dates. Nodes with *asterisks* indicate inferred transitions to the underground habitat (adapted from Emerling and Springer (2014), paintings by Michelle S. Fabros, redrawn by Monika Hänel)

subterranean environments. The incidence of regressive mutations in retinal genes (pseudogenization) post-dates the inferred entrances into their subterranean habitat and increases with time (Emerling and Springer 2014).

Rudimentation is not only restricted to morphological traits. Noctuid moths endemic to the mountains of the South Pacific island of Tahiti have evolved in an environment without bats and have lost their defensive behaviour against these predators, the acoustic startle response (Fullard et al. 2007). This response is activated by a single receptor neuron, which still exists, but exhibits a rudimentary reduced firing activity. Partial regression in the nervous control of a defensive behaviour is also observed in the Pacific field cricket *Teleogryllus oceanicus* (Orthoptera, Gryllidae), whose distribution ranges naturally across the South Pacific Ocean from Indonesia to French Polynesia, in those areas characterized by the absence of bats (Fullard et al. 2010).

Another example is provided by the penguins, all of which have lost the sweet, umami, and bitter tastes and only the genes responsible for salty and sour are intact (Zhao et al. 2015). It is suggested that a key component of taste transduction in the gustatory system of bitter, sweet, and umami but not of sour or salty taste, the cation channel *Trpm5*, is temperature sensitive, with lower activities at lower temperatures. It may have been effectively non-functional in the taste buds of ancestral penguins, which radiated after the formation of the large Antarctic ice sheet about 30 million years ago, rendering the tastes relying on this channel unusable and gradually lost. For the giant panda, it was shown that the functional constraint of the umami taste receptor gene *Tas1r1* relaxed leading to pseudogenization at about 4.2 Ma with its dietary switch from a carnivorous to a vegetarian diet (Zhao et al. 2010). Similarly, the pseudogenization of a sweet-receptor gene was found to account for the indifference of cats to detecting sweetness (Li et al. 2005). This molecular change detected in tigers, cheetahs, and the domestic cat was probably a loss due to the carnivorous type of feeding of Felidae, in which sweetness has no importance.

From the view of a human being relying on sight as the dominant sense, the most bizarre and striking examples for rudimentation—often also characterized as degeneration or regression of traits—are provided by the most conspicuous loss of eyes or dark pigmentation in species living in the continuous absolute darkness of subterranean habitats like caves (Juberthie 2000; Juberthie and Decu 1994; Culver and Wilkens 2000). Such so-called troglobites are found in almost every taxonomic group distributed over all continents except Antarctica. The outer appearance of these blind and white animals seems mysterious and has fascinated scientists for a long time. Even Darwin found explaining regressive evolution difficult.

Such continuously dark habitats are usually denominated as caves, which are big enough for humans to enter. However, these are not the only places where cave species may evolve. Smaller animals inhabit narrower fissures; for example, those found in lava rocks in large numbers. The majority of such habitats originated by hydrological erosion in karstic limestone (Culver and Pipan 2009; Mitchell et al. 1977) or formed in the lava fields of volcanic eruptions (Juan et al. 2010). Caves may even exist at the bottom of the oceans as demonstrated by some marine lava tubes (Wilkens et al. 2009). A specific type of cave exists close to marine coasts and is termed anchialine (Iliffe 2000; Pérez-Moreno et al. 2016). Such habitats are landlocked with a subterranean connection to the sea and fresh and salt water may be stratified. They are colonized by marine cave animals. The cave species living here may actively invade the inland freshwater parts of such habitats or, after the

**Fig. 2.7** The night-active troglophile surface catfish *Rhamdia laticauda* (**a**), and specimens of its eye- and pigment-reduced cave sister species *R. zongolicensis*. Cave specimens kept in permanent darkness (**b**) or in light (**c**), when dark pigmentation reappears

sea has regressed, adapt to live in the fresh water now filling the former anchialine caves. Fossil coast lines of no-longer extant oceans are still detectable today because of the geographical position of such caves.

Cave animals are completely cut off from light as a mode of energy and information. Unlike in fossorial mammals, cave animal evolution proceeds in continuous absolute darkness and is for several reasons extraordinary. In addition to the regression of eyes and black body pigmentation, many other traits including behavioural ones may get reduced. Simultaneously, adaptive traits are constructively improved.

In general, the common ancestors of cave species are night-active forms, so-called troglophiles, which are preadapted to cave life because of already having acquired and improved traits necessary to live in light-poor environments or to be active at night. For example, whereas among teleosts the diurnal characid fish (Characiformes) are only represented by three cave species, the mostly nocturnal catfish (Siluriformes) contain at least one third of the 151 cave fish species in total known today (Fig. 2.7) (Proudlove 2010). For such troglophiles, life in light-poor or lightless environments is not extreme. They exhibit morphological, physiological, and behavioural adaptations by which they are able to find food and propagate. They are even active cave invaders. Thus, the process of rudimentation may start immediately after cave colonization and may proceed comparatively quickly because there is no need for constructive functional compensation. In cave colonizing species, usually no gradual substitution of function exists as, for example, can be assumed for locomotion between the regressing hind legs and the constructive tail fin in the evolution of whales. This is an almost unique characteristic of cave animals and might accelerate the tempo of eye or pigment regression in them. This may explain why cave descendants and their surface ancestors are often closely related sister species or forms that are still interfertile.

Cave-living species have evolved in almost every systematic group of animals and in almost every type of lightless habitat (Culver and Pipan 2009; Juberthie and Decu 1994; Culver and Wilkens 2000). They occur in marine or fresh water as well as in terrestrial environments. The only common abiotic factor they are submitted to is continuous absolute darkness. In the temperate zones, low energy supply very often plays a role as a second cave characteristic, but is often overestimated as an important requisite in the evolution of cave animals because many caves (in particular, tropical ones) are, as a rule, food abundant (Gnaspini and Trajano 2000; Fernandes et al. 2016).

## References

Abel O (1908) Die Morphologie der Hüftbeinrudimente der Cetaceen. LXXXI. Band der Denkschrift der Mathematisch-Naturwissenschaftlichen Klasse der Kaiserlichen Akademie der Wissenschaften, vol 81, pp 138–195

Crish SD, Dengler-Crish CM, Catania KC (2006) Central visual system of the naked mole-rat (*Heterocephalus glaber*). Anat Rec A Discov Mol Cell Evol Biol 288(2):205–212

Culver DC, Pipan T (2009) The biology of caves and other subterranean habitats. Oxford University Press, New York

Culver DC, Wilkens H (2000) Critical review of the relevant theories of the evolution of subterranean animals. In: Wilkens H, Culver DC, Humphreys WF (eds) Ecosystems of the world: subterranean ecosystems, vol 30. Elsevier, Amsterdam, pp 381–398

Deimer P (1977) Der rudimentäre Extremitätengürtel des Pottwals (*Physeter macrocephalus* Linnaeus, 1758), seine Variabilität und Wachstumsallometrie. Z Säugetierkd 42:88–101

Emerling CA, Springer MS (2014) Eyes underground: regression of visual protein networks in subterranean mammals. Mol Phylogenet Evol 78:260–270

Fernandes CS, Batalha MA, Bichuette ME (2016) Does the cave environment reduce functional diversity? PLoS One 11(3):e0151958. doi:10.1371/journal.pone.0151958

Fullard JH, Ratcliffe JM, ter Hofstede H (2007) Neural evolution in the bat-free habitat of Tahiti: partial regression in an anti-predator auditory system. Biol Lett 3:26–28

Fullard JH, ter Hofstede HM, Ratcliffe JM et al (2010) Release from bats: genetic distance and sensoribehavioural regression in the Pacific field cricket, *Teleogryllus oceanicus*. Naturwissenschaften 97:53–61. doi:10.1007/s00114-009-0610-1

Gnaspini P, Trajano E (2000) Guano communities in tropical caves. In: Wilkens H, Culver DC, Humphreys WF (eds) Ecosystems of the world: subterranean ecosystems, vol 30. Elsevier, Amsterdam, pp 251–268

Iliffe TM (2000) Anchialine cave ecology. In: Wilkens H, Culver DC, Humphreys WF (eds) Ecosystems of the world: subterranean ecosystems, vol 30. Elsevier, Amsterdam, pp 59–76

Juan C, Guzik MT, Jaume D et al (2010) Evolution in caves: Darwin's 'wrecks of ancient life' in the molecular era. Mol Ecol 19:3865–3880

Juberthie C (2000) Generalities and the diversity of the karstic and pseudokarstic hypogean habitats in the world. In: Wilkens H, Culver DC, Humphreys WF (eds) Ecosystems of the world: subterranean ecosystems, vol 30. Elsevier, Amsterdam, pp 17–39

Juberthie C, Decu V (eds) (1994) Encyclopaedia Biospeologica. I. Société de Biospéologie. Moulis, France

Li X, Li W, Wang H et al (2005) Pseudogenization of a sweet-receptor gene accounts for cats' indifference toward sugar. PLoS Genet 1(1):27–35

Mitchell RW, Russell WH, Elliott WR (1977) Mexican eyeless characin fishes, genus *Astyanax*: environment, distribution, and evolution, vol 12. Texas Tech Press, Lubbock, pp 1–89. Special publications of the Museum Texas Tech University

Mitchell KJ, Llamas B, Soubrier J et al (2014) Ancient DNA reveals elephant birds and kiwi are sister taxa and clarifies ratite bird evolution. Science 344:898–900

Nikitina NV, Maughan-Brown B, O'Riain MJ et al (2004) Postnatal development of the eye in the naked mole rat (*Heterocephalus glaber*). Anat Rec A Discov Mol Cell Evol Biol 277 (2):317–337

Pérez-Moreno JL, Iliffe TM, Bracken-Grissom HD (2016) Life in the underworld: Anchialine cave biology in the era of speleogenomics. Int J Speleol 45(2):149–170

Phillips MJ, Gibb GC, Crimp EA et al (2010) Tinamous and moa flock together: mitochondrial genome sequence analysis reveals independent losses of flight among ratites. Syst Biol 59 (1):90–107

Proudlove GS (2010) Biodiversity and distribution of the subterranean fishes of the world. In: Trajano E, Bicuette ME, Kapoor BG (eds) Biology of subterranean fishes. Science Publishers, New Hampshire, pp 65–80

Thewissen JGM, Bajpai S (2001) Whale origins as a poster child of macroevolution. BioScience 51(12):1037–1049

Wilkens H, Iliffe TM, Oromí P et al (2009) The Corona lava tube, Lanzarote: geology, habitat diversity and biogeography. Mar Biodivers 39:155–167. doi:10.1007/s12526-009-0019-2

Zhao H, Yang J-R, Xu H et al (2010) Pseudogenization of the umami taste receptor gene *Tas1r1* in the giant panda coincided with its dietary switch to bamboo. Mol Biol Evol 27(12):2669–2673. doi:10.1093/molbev/msq153

Zhao H, Li J, Jianzhi Z (2015) Molecular evidence for the loss of three basic tastes in penguins. Curr Biol 25(4):141–142

# 3 Diversity and the Phylogenetic Age of Cave Species


**Abstract**

A hot spot of biodiversity of aquatic cave species of very different biogeographic and systematic origins is located around the Gulf of Mexico. In continental Mexico, Neotropic and Nearctic cave faunas overlap north as well as south of the trans-Mexican volcanic belt and may even co-occur in the same cave. These species may also be associated with marine relics, closely related species that inhabit caves of peri-Carribean islands like Cuba and the Bahamas or of the Yucatán peninsula. Other cave species occurring in this area derive from deep sea forms or show close relationships to species occurring in subterranean water on the volcanic Eastern Atlantic island of Lanzarote. Representatives of the thermosbaenaceans and the remipedes, the latter having developed their centre of radiation in the peri-Carribean Islands, exclusively live in caves but exhibit circumtropical Tethyan distribution.


The distribution and the degree of eye reduction of cave species mirror past geological events and, in particular, Pleistocene climatic changes. Remipedes and thermosbaenaceans represent the oldest and completely eye-reduced cave inhabitants. They started evolution in the darkness of caves along Tethyan shorelines. Another group of strongly eye-reduced species including bythitid cusk eels, diverse shrimps, mysids, and cirolanid isopodes derive from marine ancestry inhabiting caves since the end of the Pliocene. Other species are characterized by less strong degrees of eye reduction. They are represented by Nearctic crayfish, Neotropic hepapterid catfish and characids, bythitid cusk eels, and several shrimp species. Their occurrence is concentrated south of the trans-Mexican volcanic belt (TMVB) and in the Bahamas. Because of the various effects of climatic cooling, they probably colonized caves as late as after the last glacial maximum during the Wisconsin glaciation 27,000–24,000 years ago, or in part even after the Dryas period, which ended about 12,000 years ago.



H. Wilkens, U. Strecker, *Evolution in the Dark*, DOI 10.1007/978-3-662-54512-6_3

## 3.1 Mexico and the Peri-Caribbean Islands: A Hot Spot of Aquatic Cave Species Diversity

Besides paleness caused by the loss of dark pigments, eye reduction is the most conspicuous trait of cave-living animals. It is notable that its degree may vary and all stages between the initial slight reduction in size and a finally no longer externally visible tiny rudiment, which in fish and amphibia is deeply sunken into the interior of the head and overgrown by tissue, can be found. It has long since been in the centre of interest to determine how long it takes for an eye to become reduced. Modern molecular genetic methods have provided important results. The outcomes are often hampered, though, by the uncertainties arising from calibrating the molecular clock and dealing with rather large ranges of time. Furthermore, another point of debate is whether the date found actually corresponds to that of cave entry or just reveals the splitting of ancestral surface lineages, with one becoming the ancestor of a cave form at a later date. This would give the impression of a much longer period than it would actually take for regression. An alternative method to determine the time necessary for reduction is to directly involve the degree of trait regression based on the assumption that the state of regression is correlated with the time since evolution in absolute darkness began. The earlier it started and the more time that has passed, the higher the degree of regression is expected to be. As the eye, in contrast to dark pigmentation, is not or only to a small extent submitted to environmental modification by missing light stimuli in darkness, it is suitable for this purpose. The distribution and the degree of eye reduction of cave species may be correlated with and mirror past geological events and climatic changes (Humphreys 2000; Kosswig 1967; Wilkens 1986; Wilkens et al. 1989).

The most suitable region for such studies is situated around the Gulf of Mexico in Florida and in Mexico, including the pensinsula of Yucatán as well as the adjacent Peri-Caribbean Islands, in particular the Bahamas and Cuba (Fig. 3.1). This area is a hot spot for the diverse origins of aquatic cave species belonging to different taxonomic groups and biogeographic regions, which often coexist in the same cave. There are several reasons for this complexity:

- In Mexico, the volcanic eruption zone of the TMVB separates the biogeographical transition area between Neotropis and Nearctis (Gómez-Tuena et al. 2007). This is not only reflected by surface animal and plant species but also by the cave fauna (Bussing 1985). In caves located at the eastern margins of the Sierra Madre Oriental and the Sierra Madre del Sur, which are separated by the TMVB, in the Chiapas/Guatemala Highlands (Central Chiapas Highlands), and in the Maya Mountains in Belize, cave species of diverse taxonomic groups exist that derive from Nearctic North as well as from Neotropic South American freshwater ancestors.
- Some of the cave fauna occurring at the eastern margin of the Sierra Madre Oriental north of the TMVB have marine origin. They are closely related to freshwater cave species of the Peri-Caribbean islands and the Mexican

**Fig. 3.1** The Upper Central American Nearctic and Neotropic transition zone, the Peninsula of Yucatán, and the Peri-Caribbean Islands (Cuba, Bahamas) contain important cave areas. *SMO* foothills of Sierra Madre Oriental, *SMO* eastern foothills of Sierra Madre del Sur, *SMC* Sierra Madre de Chiapas, *TMVB* Trans-Mexican Volcanic Belt, *MM* Maya Mountains

pensinsula of Yucatán (Iliffe and Botoşăneanu 2006). These faunas originally evolved in marine anchialine caves situated on ancient Pliocene coasts, which have become freshwater caves after the sea retreated.

- In the Peri-Caribbean Islands and the Yucatán, marine fauna exists that is related to cave species occurring in other regions of the world. Their closest taxonomic relatives occur on the Canary Island of Lanzarote in the Eastern Atlantic and a few even live in Western Australian caves. Such species exhibit so-called amphiatlantic or even circumtropical distribution, respectively (Wilkens et al. 2009).
- In addition, part of the species assemblage of the Peri-Caribbean Islands and the Yucatán derive from deep sea ancestors.

In these areas, cave species exhibiting all stages of eye reduction and reduction of dark pigments exist. They can be separated into two main groups, those showing strongly reduced eyes on the one hand (the phylogenetically older species), and the phylogenetically younger ones that are just starting eye reduction or exhibiting a more or less intermediate position with respect to the ancestral form.

For the first group, it is impossible to determine when the process of eye regression reached its recent state of extreme reduction, which may have been a long time ago. Therefore, all we know is when a cave species began to inhabit caves and how old it is in geological terms. In the second group, the process of eye reduction has not yet finished and we can determine how much time the evolutionary process of reduction of a trait like the eye takes, provided geological or climatic dates allow the determination of its starting point.

## 3.2 Strongly Eye-Reduced Cave Species

In the crustaceans belonging to this taxonomically very diverse group, the eye optical apparatus consisting of crystalline cones, ommatidia, and visual ganglia is completely reduced. They include relic crustaceans like members of the class Remipedia and the pancarid Thermosbaenaceans (Hendrickson et al. 2001; Koenemann and Iliffe 2013; Mejía-Ortíz 2005; Stegner et al. 2015; Wilkens 1972; Yager 1981) and exclusively occur in marine caves in circumtropical distribution (Fig. 3.2) (Iliffe et al. 1984; Wilkens et al. 2009). It is assumed that they descend from some widespread and disjunct marine ancestors in the Mesozoic Tethys Ocean (Iliffe and Botoşăneanu 2006; Neiber et al. 2011; Olesen et al. 2014). In accordance with this, it can be assumed that these species have inhabited caves for a very long time, maybe about 120 million years. During this period, the remipedes were even subjected to a process of adaptive radiation within the caves of the Bahamas, resulting in 13 out of 27 extant species (Hoenemann et al. 2013; Jaume et al. 2013; Koenemann and Iliffe 2013; Neiber et al. 2011; Yager 1981).

Another group of marine cave species derive from deep sea ancestors. Some of them are distributed on both sides of the Atlantic. Most probably these forms had already reduced eyes when colonizing the limestone or volcanic caves of the Bahamas and Lanzarote, respectively (Jaume et al. 2013; Wilkens et al. 2009). For example, the pardaliscid amphipods *Speleonicippe buchi* (Fig. 3.3) and *S. provo* occur in Lanzarote or the Bahamas, respectively. The same is valid for the macellicephaline polychaetes *Gesiella jameensis* (Fig. 3.4) and *Pelagomacellicephala iliffei*.

In contrast, a group of freshwater crustacean and fish species began cave life at a more recent date in marine anchialine caves located on Pliocene coasts north of the TMVB at the foothills of the Mexican Sierra Madre Oriental along the Tampico

**Fig. 3.2** The remipede *Morlockia ondinae* (3.0 cm body length) was found in a submerged lava tube in Lanzarote (Canary Islands). This group has its closest relatives in Peri-Caribbean and Yucatán marine caves (Photo Ulrike Strecker)

**Fig. 3.3** The blind cave amphipod *Speleonicippe buchi* (Pardaliscidae) (1.0 cm) from marine groundwater in Lanzarote (Canary Islands) derives from deep sea species (Photo Ulrike Strecker)

**Fig. 3.4** The polychaete *Gesiella jameensis* (Macellicephalinae) (0.8 cm) occurs in a lava tube in Lanzarote (Canary Islands) and is closely related to deep sea species (Photo Ulrike Strecker)

embayment and in the Yucatán as well as in Cuba (Fig. 3.1) (Bauer-Gottwein et al. 2011; Iturralde-Vinent et al. 2016; Weyl 1964; Wilhelm and Ewing 1972). All of them became freshwater species after the sea retreated from the Pliocene coastlines before the beginning of the Pleistocene (Beddows 2003). They include several bythitid cusk eel species of the genus *Lucifuga* from Cuba and the species *Typhliasina pearsei* from the Yucatán (Fig. 3.5) (Hunter et al. 2007), the widely distributed species-rich cirolanid isopodes (Fig. 3.6) (Iliffe and Botoşăneanu 2006), palaemonid shrimps of the genera *Troglocubanus* (Cuba) and *Troglomexicanus* (Mexico), as well as both a palaemonid (*Creaseria morleyi*) (Figs 3.7 and 3.12) (Botello and Alvarez 2010) and several atyid shrimps of the genus *Typhlatya,* and also mysids (Hunter et al. 2007).

In caves at the foothills of the Sierra Madre Oriental del Norte, species deriving from freshwater ancestors co-occur with the before-mentioned species of marine origin. These are a series of cave populations of the Neotropic characid fish *Astyanax* (Fig. 3.8), which as a salt-intolerant primary freshwater fish could only reach North America after the Central American land bridge had formed in the late Pliocene, about 3 million years ago (mya) (see Sect. 4.3.1). Furthermore, Nearctic ictalurid cave catfish, genus *Prietella* (Fig. 3.9), and the decapod crayfish *Procambarus xilitla* are living here. It is hypothesized that these freshwater species were able to colonize the former anchialine caves after they had become freshwater habitats at the end of the Pliocene.

**Fig. 3.5** The cave cusk eels *Typhliasina pearsei* (**a**) from Yucatán and several species of the genus *Lucifuga* (**b**) from Cuba are widespread in the underground freshwater systems of these areas. These fish are live-bearing and the belly of the *T. pearsei* female is protruding because of being pregnant with two juveniles (Photo of *Lucifuga* J. P. Durand)

**Fig. 3.6** The isopod *Speocirolana palaezi* (Cirolanidae) is about 1 cm in size and occurs in the Sierra de El Abra (Sierra Madre Oriental, Northeastern Mexico)

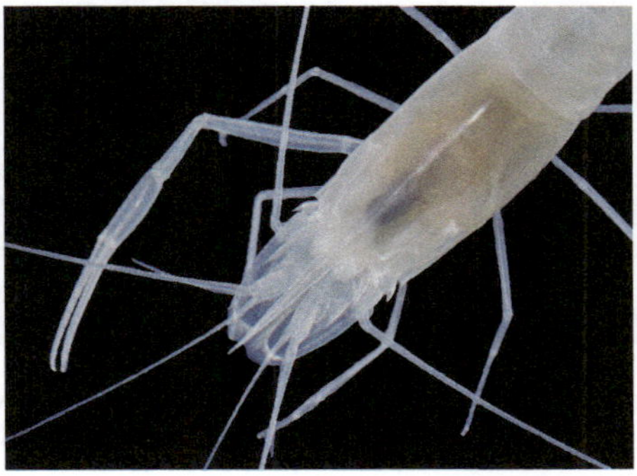

**Fig. 3.7** The eye stalks are strongly reduced and dark pigmentation is completely missing in the cave shrimp *Creaseria morleyi* (Palaemonidae, up to 6 cm long) from Yucatán underground freshwaters. Its capillary first antennae are extremely prolonged to 8-fold the length of the body

**Fig. 3.8** A cave characid, genus *Astyanax*, from Cueva de El Pachón in the Sierra de El Abra (Sierra Madre Oriental, Northeastern Mexico)

**Fig. 3.9** *Prietella phreatophila* (Ictaluridae) is one of three Nearctic cave catfish species occurring in caves along the Sierra Madre Oriental in Northeastern Mexico (Photo Dean Hendrickson)

## 3.3 Lesser and Variably Eye-Reduced Cave Species

The only representatives of this group occurring in the cave area of the Sierra Madre Oriental north of the TMVB are some of the cave populations of the Neotropic freshwater characid fish *Astyanax* (see Sect. 4.2). In striking contrast to this, the caves south of the TMVB at the eastern margin of the Sierra Madre del Sur and in the Central Chiapas Highlands are almost exclusively inhabited by species with eyes of intermediate reduction (Fig. 3.1). This assemblage is composed of Nearctic cambarid crayfish, Neotropic hepapterid catfish, and freshwater shrimps. Several species and subspecies of the decapod crayfish genus *Procambarus* (Fig. 3.10) and the shrimps *Macrobrachium villalobosi, M. sbordonii*, and *Cryphiops luscus* (Palaemonidae), as well as the only freshwater alpheid shrimp *Potamalpheops stygicola* (Alpheidae), some of which even co-occur in the same caves (Fig. 3.11), exhibit a comparable degree of eye reduction (Mejía-Ortíz 2005; Mejía-Ortíz et al. 2003, 2008). All crustacean species possess well developed ommatidia with crystalline cones, but in very reduced number. Correlated to this, the three optic ganglia of the crustacean eye stalk are reduced in size, too (Wilkens 1986) (Fig. 3.12). Also, several cave catfish species of the Neotropic genus *Rhamdia* (Hepapteridae) co-occur with the cave *Procambarus* crayfish species (Fig. 2.7). They are sister species of the surface fish *R. laticauda*, which, as a

**Fig. 3.10** The cambarid cave crayfish *Procambarus oaxacae reddelli* inhabits five geographically separated caves south of the Trans-Mexican Volcanic Belt (rostrum-telson length 6 cm). Insert shows reduced number of corneal facets and ommatidia at the tip of the eye stalk

**Fig. 3.11** The cave shrimp *Macrobrachium villalobosi* (Palaemonidae) (rostrum-telson length 4 cm) carries a reduced number of corneal facets and ommatidia on the eye-stalk tip. The eye stalk is coloured red at its tip because of mutations in the ommochrome pigment. This species co-occurs with *P. oaxacae reddelli*

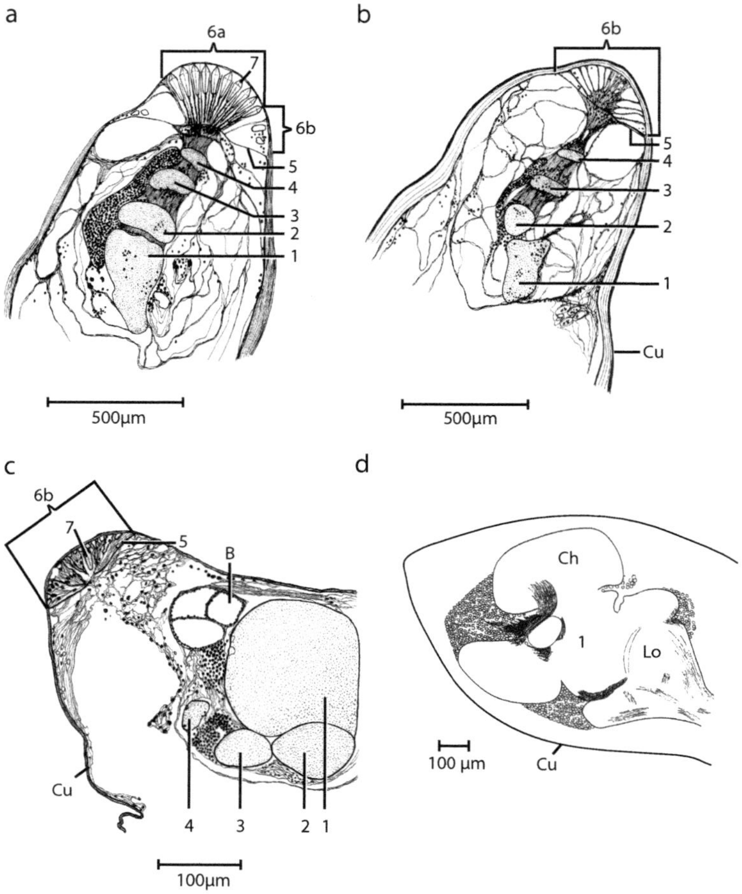


**Fig. 3.12** Reduction of the eye of decapod crustaceans is a process of size diminution of its component parts involved in vision. As compound eyes grow at their margins, the regression process is characterized by cornea facets and ommatidia shrinking in number from the periphery at the beginning. Schematic longitudinal sections of the eye stalks of the cambarid cave crayfish *Procambarus oaxaca reddelli* (**a**) from Southern Mexico, *P. erythrops* (**b**) from Florida, the cave shrimps *Typhlatya garciai* (Atyidae) (**c**) from the Bahamas, and *Creaseria morleyi* (Palaemonidae) (**d**) from Yucatán. 1 = medulla terminalis, 2 = m. interna, 3 = m. externa, 4 = lamina ganglionaris, 5 = basal membrane, 6a = number of ommatidia reduced, 6b = lenses reduced, 7 = intact lens, 8 = Bellonci organ, *Cu* cuticula (adapted from Wilkens 1973, 1986; Juberthie-Jupeau 1976)

salt-intolerant primary freshwater fish, could only reach North America after the Central American land bridge closed in the late Pliocene about 3 mya ago (Perdices et al. 2005). Except for *R. macuspanensis*, all *Rhamdia* cave species are characterized by less reduced eyes and show remarkable inter- and intraspecific variability of eye size and histological differentiation (Fig. 3.13) (Wilkens 2001).

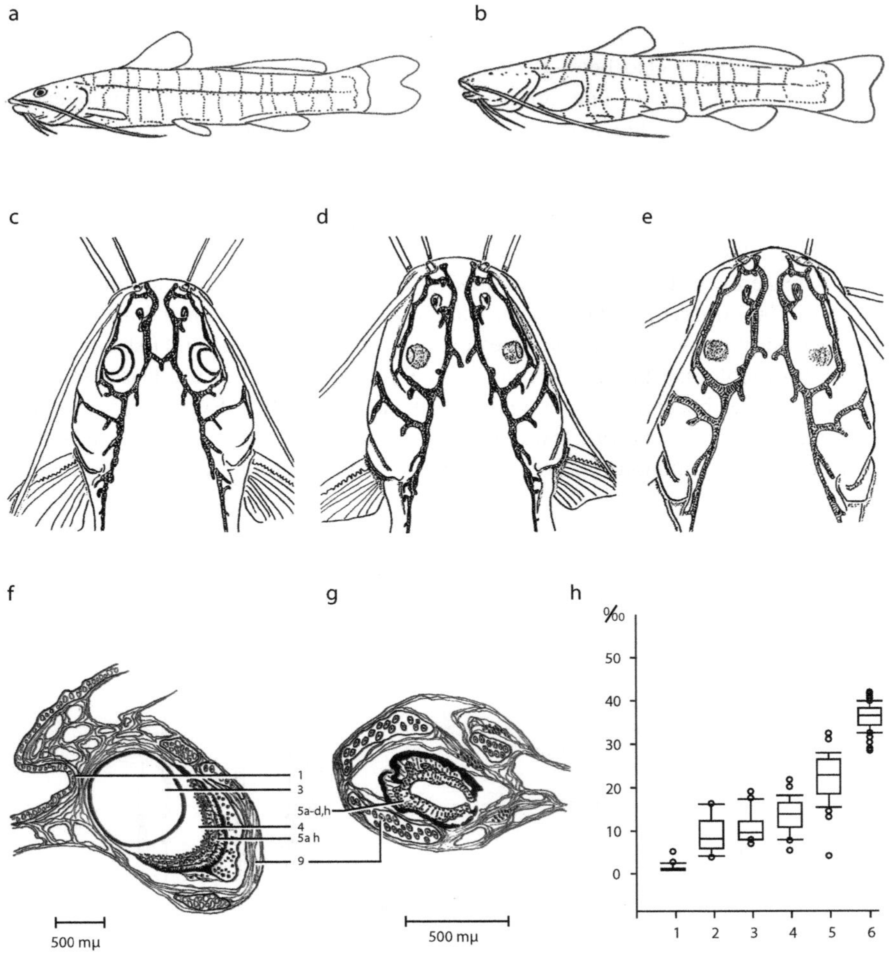


**Fig. 3.13** The large catfish genus *Rhamdia* (Hepapteridae, up to 15 cm body length) has developed at least five different cave forms, all being sister species to the troglophile surface fish *R. laticauda*. The electric sense organs, which are arranged in vertical rows along the sides of the whole body, are not improved in the cave forms *R. zongolicensis* and *reddelli* (**b**) when compared with the surface sister species *R. laticauda* (**a**). The same is valid for the head canal system (**c** = surface, **d**, **e** = cave). In contrast, the barbels are elongated in the cave fish (**b**) and the eyes have become reduced showing variability in size between the different cave species (**h**) as well as within particular species like *R. zongolicensis* (abbreviations in f and g: 1=cornea, 3=lens, 4=vitreous body, 5=retina rudiment, 9=sclera). 1 = *R. macuspanensis*, 2 = *R. laluchensis*, 3 = *R. reddelli*, 4 = *Rhamdia* undescribed species, 5 = *R. zongolicensis*, 6 = *R. laticauda* (surface species), measured in per mill of body length (adapted from Weber 2003)

The live-bearing toothcarp *Poecilia mexicana* (Poeciliidae) with highly variable but still functional eyes from a sulfidic cave located at the foothills of the Central Chiapas Highlands also belongs to this group (Fig. 3.14) (Joachim et al. 2013; Peters et al. 1973). Its late origin is corroborated by molecular phylogenetic analysis, which revealed it to be from the post-Pleistocene age (Palacios et al. 2016).

It is suggested that above all, Pleistocene climatic cooling and its concomitants are responsible for the almost exclusive existence of less eye-reduced cave species south of the TMVB in the Sierra Madre del Sur and the Central Chiapas Highlands. The water temperature of the rivers flowing from the adjacent high mountain areas of up to 2000 m feeding these caves must have been too low for the ancient cave faunas to survive. Therefore, the ancestors of the recent decapod and fish cave fauna could not recolonize these caves before the end or even after the last glaciation, when the water became warmer again (Wilkens et al. 1991). Even today, climatological frost may appear for a few days in higher altitudes and the different *Rhamdia* cave species only inhabit relatively warm caves at low ground levels at the margin

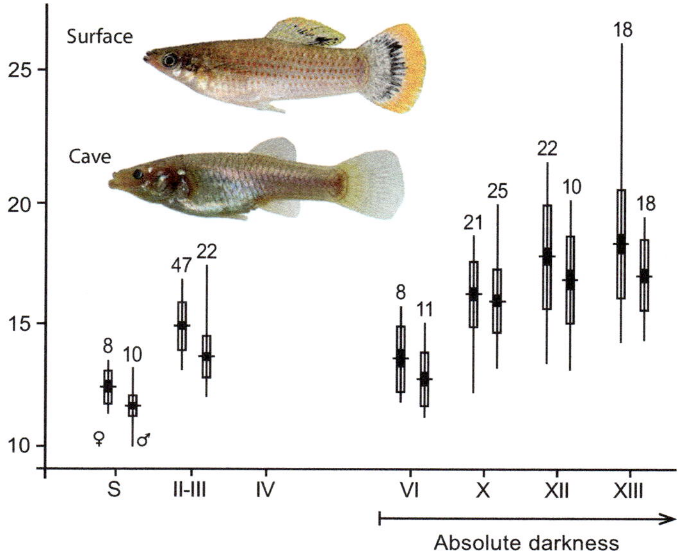


**Fig. 3.14** Gradient of relative size and variability of the eye of a cave population of *Poecilia mexicana* (Poeciliidae) compared with its surface sister form. The cave fish inhabit different cave sections from chamber I, close to the entrance, to chamber XIII, which is the deepest inside the cave. Eye size and variability are presented for male and female specimens separately for each cave section II–XIII. Locations II–IV lie under dim skylights, whereas VI–XIII represent absolutely dark sections. The female cave fish exhibit a constructively enlarged tissue pad in the ventral genital region, which serves as female chemical identification for the males in darkness (adapted from Peters et al. 1973; photos by Michael Tobler)

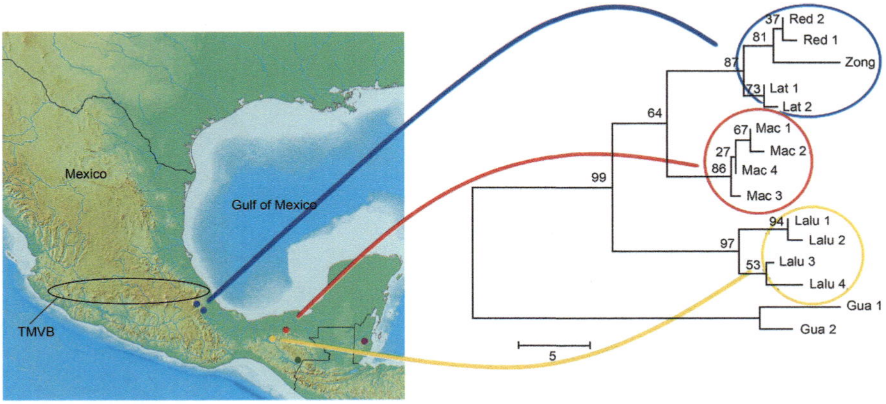


**Fig. 3.15** Phylogeography of surface *Rhamdia laticauda* (*Lat*) and its cave sister species *R. reddelli* (*Red*), *R. zongolicensis* (*Zong*), *R. macuspanensis* (*Mac*), and *R. laluchensis* (*Lalu*) based on cytochrome b sequence data. Outgroup is *R. guatemalensis* (Gua) from the Yucatán peninsula and the Río San Antonio (Oaxaca) that originates within the *R. reddelli* cave. *Green* (undescribed *Rhamdia* population) and *purple dots* (R. laticauda typhla from Belize) indicate further cave *Rhamdia* localities, (*TMVB* Trans-Mexican Volcanic Belt) (adapted from Wilkens et al. 1991; Weber et al. 2003)

of the Sierra Madre del Sur (2), the Chiapas Highlands (3), and the Maya Mountains (Fig. 3.15).

Consequently, surface ancestors and cave derivatives are still closely related. A cytochrome b study revealed the small genetic distance between surface and cave *Rhamdia*. For example, between the cave fish *R. reddelli* and the surface *R. laticauda* population living in the same drainage system, it is even smaller than it is between surface *R. laticauda* populations from different surface drainages (Fig. 3.15) (Perdices et al. 2002). The high variability of eye size found in almost all *Rhamdia* cave species, including *R. laticauda typhla* from the adjacent Belizean Maya Mountains, corroborates the relatively recent time of origin of these cave species (Greenfield et al. 1982; Wilkens 2001). Only *R. macuspanensis* does not show variable eye size and is an exception among the cave *Rhamdia* species. It may have been protected from extinction because it inhabits the most eastern cave located within an isolated mountain ridge in the warmer Gulf Coast Lowlands (Weber and Wilkens 1998) (Figs. 3.13 and 3.15).

The eyes of the bythitid cave cusk eel *Lucifuga lucayana* and *L. spelaeotes* from the Bahamas also exhibit high variability and are far less reduced than those of the closely related Cuban *Lucifuga* species, which were included in the strongly eye-reduced group (Figs. 3.16 and 3.17) (García-Machado et al. 2011; Möller et al. 2006; Wilkens 1986; Wilkens et al. 1989). Furthermore, in contrast to the *Typhlatya* shrimp species from Yucatán, the eye stalks of *T. garciai* from the Bahamas still contain rudimentary retinal cells and all optic ganglia—though

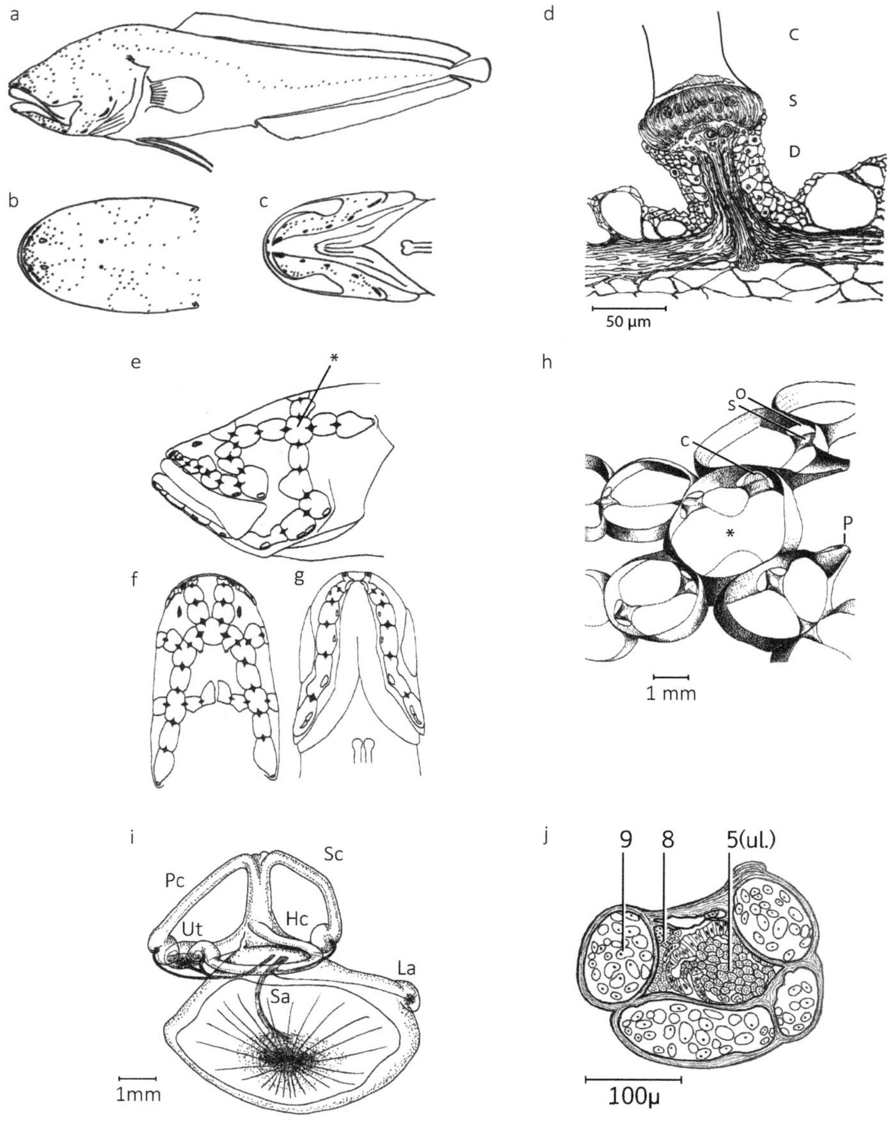


**Fig. 3.16** The cave cusk eel *Typhliasina pearsei* (Bythitidae) (**a**) from Yucatán is particularly specialized for living in calm underground waters. The head lateral line canals (**e**, **f**, **g**) are enlarged to voluminous chambers, which are separated from the outside by just very thin skin membranes. Due to this, the slightest pressure waves excited by movement of live prey can be perceived. The large sacculus (**i**) of the labyrinth organ indicates extremely good sound perception. In contrast, the number of free neuromasts (**a–c**) is rather low and the eyes are reduced to tiny rudiments (**j**). Position of free neuromasts (**a–c**), schematic histological section of a free neuromast (**d**), lateral line canal chambers on head side (**e**), above (**f**) and below head (**g**). *Asterisks* in **h** and **e** show the

these are much smaller. Except for singular ones, the crystalline cones are reduced (Fig. 3.12) (Juberthie-Jupeau 1976). The Bahama Islands are assumed to have been affected by Pleistocene glacial cooling events more than any other Peri-Carribean island and the Yucatán because of their northernmost geographic position (Fig. 3.1). Marine regressions during glacial episodes must have accentuated cooling in shallow water (Stanley and Campbell 1981; Pregill and Olson 1981). By the late Pleistocene, there apparently remained only those species that could tolerate the temperature systems of glacial stages. No terrestrial vertebrate fossil faunas have been found in the Bahamas that are older than the last (Wisconsin) glaciation. Thus, any hypothetically existing, ancient, phylogenetically older, and more strongly eye-reduced cave forms of *Lucifuga*—comparable to those occurring recently in Cuba—probably became extinct. Only those cave species that found shelter in the deeper caves and fissures of the 4000-m thick limestone block forming the Bahamas could survive. These were, for example, species with deep sea ancestry. In contrast, the surface coral reef cusk eel, genus *Ogilbia*, as well as its hypothetical ancient cave sister species, genus *Lucifuga*, became extinct. Thus, for the recent Bahamian *Lucifuga* and *Typhlatya* cave species, a very recent date of origin starting at the end of the last glaciation could be derived. In contrast, the diverse *Lucifuga* cave fish species from Cuba and *Typhliasina pearsei* as well as *Typhlatya* shrimps from the Yucatán, which originated earlier in Pliocene, could persist during this time due to their more southern distribution areas.

Among the exclusively strongly eye-reduced freshwater cave fauna of Yucatán, the cave swamp eel *Ophisternon infernale* (Synbranchidae) represents an exception. It exhibits intermediate eye reduction and after exposure to light its body pigmentation appears again to some extent and the fish become brownish (Fig. 3.18) (Wilkens 1979). Its ancestral sister form, the Neotropic secondary salt-tolerant surface freshwater fish *O. enigmaticum*, could have crossed the marine strait between North and South America as early as during the Miocene (12.7–23 mya) before the uplift of the Central American land bridge (Perdices et al. 2005). However, the low degree of eye and pigment reduction of *O. infernale* does not correspond with this early date (Fig. 3.19), at which it could already have colonized the Yucatán caves. As yet, a far more reduced cave *Ophisternon* has not been found, although such an eye-reduced population could be expected to exist because of the early invasion date from South America. It is proposed that this can also be explained by Pleistocene climatic changes that correlated with the sea level

**Fig. 3.16** (continued) same head canal chamber in different magnifications. Abbreviations in **a–h**: *c* cupula in opening between two chambers, *d* dermal protrusion, *o* cupula in opening not drawn, *p* lateral line porus, *s* sensory cells; abbreviations in (**i**): *Hc* horizontal hemicircular canal, *La* lagena, *Pc* posterior semicircular canal, *Sa* saccule with sagitta, *Sc* superior semicircular canal, *Ut* utricle; abbreviations in **j**: 5 = vitreous body and retina rudiment (unlayered, ul.), 8 = pigmentary epithelium, 9 = cartilaginous sclera (adapted from Wilkens 1979; Schemmel 1977)

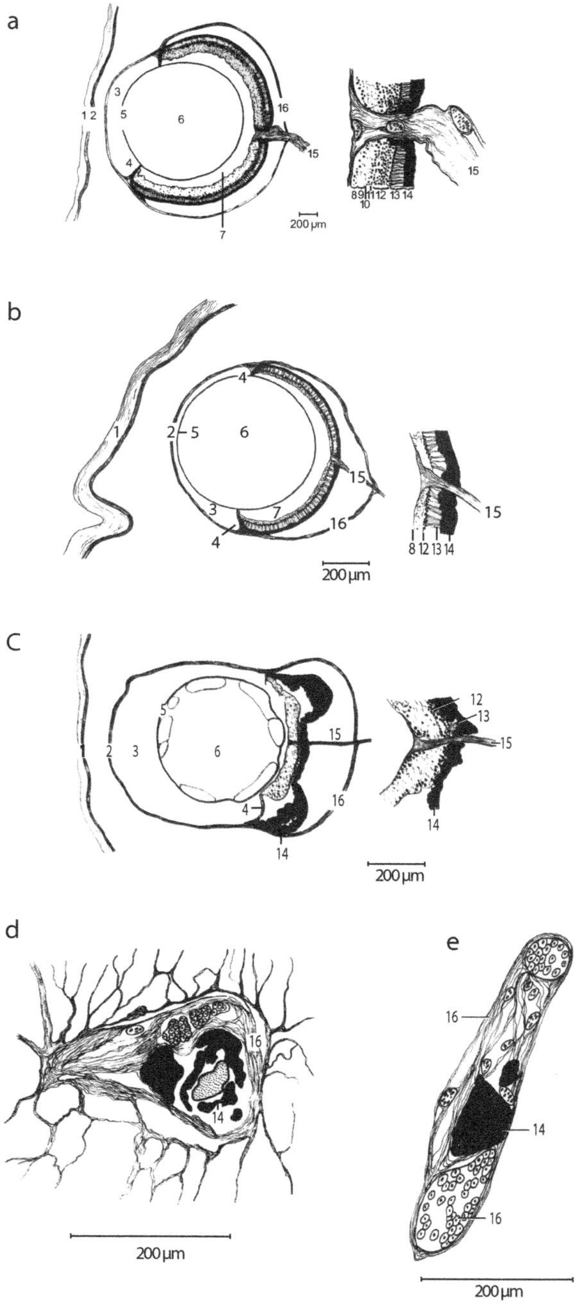


**Fig. 3.17** The eye rudiments of the surface cusk eel *Ogilbia cayorum* (**a**) and cave cusk eels from the Bahamas (*Lucifuga speleotes*, **b** and **c**), from Cuba (*Lucifuga dentatus*, **d**), and from Yucatán (*Typhliasina pearsei*, **e**). 1 = cornea (epithelial), 2 = cornea (scleral), 3 = anterior eye chamber, 4 = iris, 5 = lens capsule, 6 = lens, 7 = vitreous body, 8 = ganglionic layer, 9 = inner plexiform layer, 10 = inner nuclear layer, 11 = outer plexiform layer, 12 = outer nuclear layer, 13 = visual cell outer segments, 14 = pigmentary epithelium, 15 = optic nerve, 16 = sclera, insert showing enlarged emergence area of optic nerve (adapted from Wilkens et al. 1989)

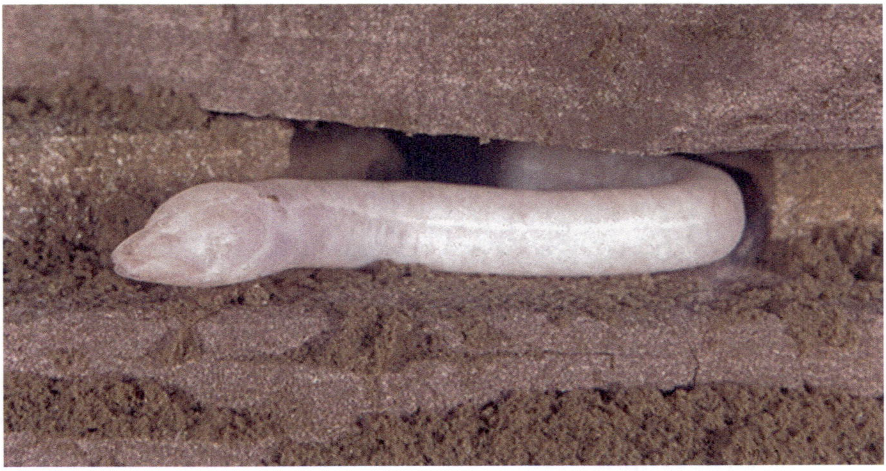

**Fig. 3.18** The swamp eel *Ophisternon infernale* (Synbranchidae) may grow up to >30 cm body length and is widely spread in the underground freshwater cave systems of Yucatán

being lowered. Because of this, more ancient hypothetical cave specimens of *Ophisternon* may have become extinct. This can be confirmed by studies of the Yucatán cave shrimp *Creaseria morleyi*, which revealed that it was submitted to population bottlenecks resulting from severe fluctuations of sea level in the Yucatán during the Pleistocene (Botello and Alvarez 2010). Possibly, the present population of the blind swamp eel *O. infernale* only started troglobitic evolution when the sea began to rise again, reaching its present day height 5000 years ago (Botello and Alvarez 2010).

A further example of a relatively recent cave invasion is provided by the decapod cave crayfish species of Florida (*Procambarus erythrops*, *P. lucifugus alachua*, *P. l. lucifugus*, *P. orcinus*, and *Cambarus cryptodytes*). They uniformly exhibit slightly stronger eye reduction than the crustaceans distributed south of the TMVB and have lost the crystalline cones completely. No proper rhabdomeric structures are developed, but all optic ganglia, though much smaller, are still extant (Wilkens 1986) (Fig. 3.12). It is suggested that these species were able to invade the caves some time after the end of the last glacial maximum because much of Florida was subjected to sinking sea levels, which lowered inland freshwater tables with the caves falling dry and with drifting sand dunes as a result of aridity covering the peninsula.

Summarizing the biogeographical and eye histological data, it is suggested that the troglobitic evolution of the rather heterogeneous group of freshwater cave species with only slight or variable eye rudiments of an intermediate stage of regression is strongly influenced by Pleistocene climatic change, which involved

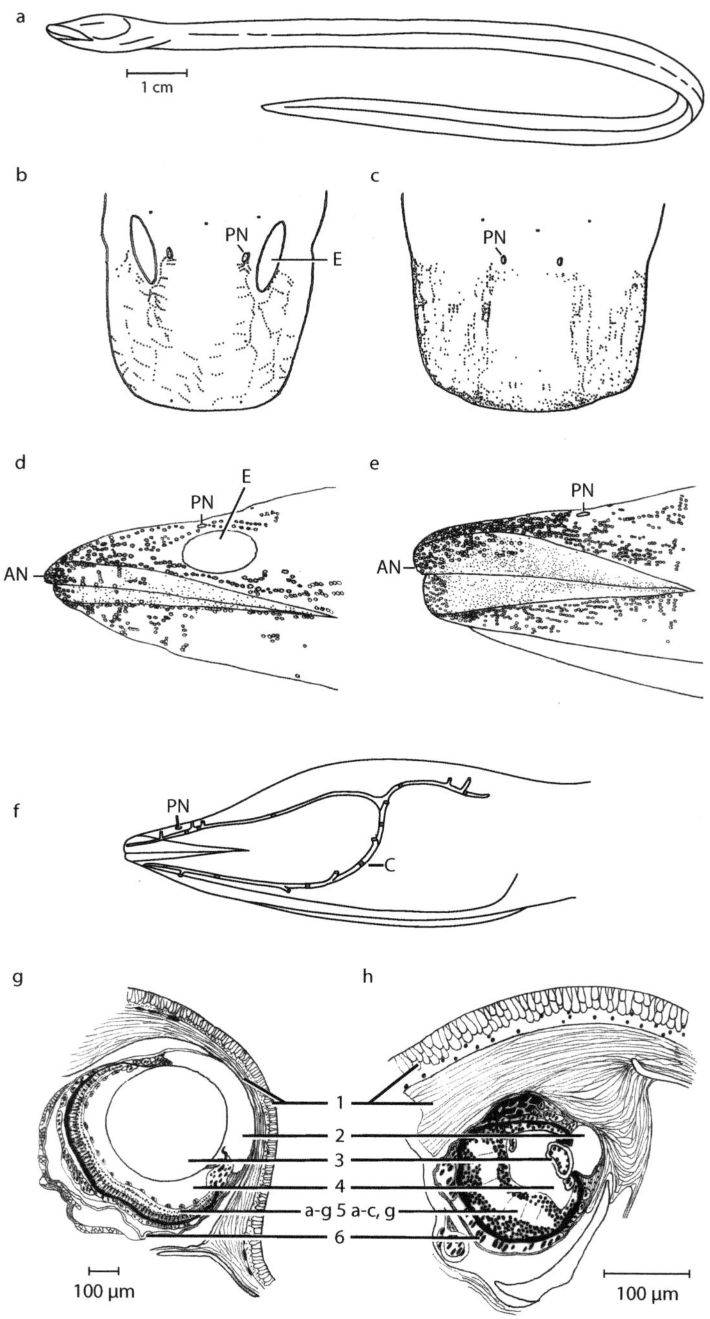


**Fig. 3.19** In comparison with the closely related troglophile surface swamp eel *Ophisternon enigmaticum* (Synbranchidae), the numbers of free neuromasts (*small dots* in **c** versus **b**, *large dots* in **e** versus **d**) are enhanced. Also, the number of taste buds (*small dots* in **e** versus **d**) is larger

**Fig. 3.20** Tiny eye stalks and red body colour of the Cuban marine shrimp *Barbouria cubensis* (Hippolytidae) indicate that this species is a troglophile occurring in marine disphotic and cave environments

temperature, precipitation, and sea level fluctuations. They may have started evolving at the earliest after the last glacial maximum during Wisconsin (27.000–24.000 BP) or even after the Younger Dryas (12,900–11,700 BP) (Clark et al. 2009; Kindler et al. 2014; Rasmussen et al. 2016).

Timing based on eye reduction will provide false results in species occurring in marine caves that either derive from deep ancestors or are distributed in the so-called disphotic or twilight zone of the ocean as well as in caves, although still able to perceive light. Such troglophile species may still possess functioning eyes, which mainly serve as light detectors so they can remain in these twilight zones and hide away from predators during daytime. The eyes may still exist, but are small in comparison with fully-eyed diurnal crustaceans or exhibit specific morphological adaptations (Frank et al. 2012; Chamberlain 2000) (Fig. 3.20). For example, in the galatheid crab *Munidopsis polymorpha* found in a lava tube and lava fissures filled with marine groundwater in Lanzarote, the eye is light sensitive but, because of the

**Fig. 3.19** (continued) because head and lips are broadened in the Yucatán cavefish *O. infernale* (**a**). The head canal system (**f**) is unchanged, but the eyes are reduced in *O. infernale* (**h**) in comparison with the surface *O. enigmaticum* (**g**). *AN*/*PN* anterior/posterior nasal opening, *C* one canal neuromast between every two pores, *E* eye. Abbreviations in **g–h**: 1 = epithelial and scleral part of cornea, 2 = anterior eye chamber, 3 = lens/lens capsule, 4 = vitreous body, 5 = (a) ganglionic layer, (b) inner plexiform layer, (c) inner nuclear layer, (d) outer plexifom layer, (e) outer nuclear layer, (f) outer segments, (g) pigmentary epithelium, (h) chorioid, 6 = sclera (adapted from Parzefall and Wilkens 1972)

**Fig. 3.21** The crab *Munidopsis polymorpha* (Galatheidae) is the only member of this species-rich deep-sea genus occurring outside the deep sea in the marine groundwater of Lanzarote (Canary Islands). Its eyes are reduced and non-imaging but nonetheless light sensitive (Photo Ulrike Strecker)

reduction of the dioptric apparatus, is no longer imaging (Fig. 3.21) (Harms 1921; Wilkens and Parzefall 1974; Wilkens et al. 2009). As *M. polymorpha* inhabits anchialine pools (in addition to the superficial lava voids of the island of Lanzarote) that receive daylight, it seeks shelter at daytime. Different eye morphology can also be found in the mysid *Heteromysoides cotti*, which occurs in the same places as *M. polymorpha* (Fig. 3.22). Its eye is small in size and exhibits variable numbers of ommatidia. It is coloured red due to mutation of ommochrome (Meyer-Rochow and Juberthie-Jupeau 1987). As in *M. polymorpha*, this eye has lost its imaging ability. It is regressive, but can distinguish different light intensity and direction.

**Fig. 3.22** The ~5 mm small mysid *Heteromysoides cotti* (Mysidacea) lives in the marine groundwater of Lanzarote (Canary Islands) and has its closest relatives in Cuban caves. Its eye pigmentation is red due to mutation of the ommochrome pigment (Photo Ulrike Strecker)

## References


Bauer-Gottwein P, Gondwe BRN, Charvet G et al (2011) Review: the Yucatán Peninsula karst aquifer, Mexico. Hydrogeol J 19:507–524

Beddows PA (2003) Cave hydrology of the Caribbean Yucatan Coast, Bulletin 11. Association of Mexican Cave Studies, Houston, USA, pp 1–96

Botello A, Alvarez F (2010) Genetic variation in the stygobitic shrimp *Creaseria morleyi* (Decapoda: Palaemonidae), evidence of bottlenecks and re-invasions in the Yucatan Peninsula. Biol J Linn Soc 99:315–325

Botello A, Iliffe TM, Alvarez F et al (2013) Historical biogeography and phylogeny of *Typhlatya* cave shrimps (Decapoda: Atyidae) based on mitochondrial and nuclear data. J Biogeogr 40 (3):594–607. doi:10.1111/jbi.12020

Bussing WA (1985) Patterns of distribution of the Central American Ichthyofauna. In: Stehli FG, Webb SD (eds) The great American biotic interchange, vol 4. Plenum Press, New York, pp 453–473

Chamberlain SC (2000) Vision in hydrothermal vent shrimp. Philos Trans R Soc Lond B 355:1151–1154

Clark PU, Dyke AS, Shakun JD et al (2009) The last glacial maximum. Science 325:710–714

Frank MF, Johnsen S, Cronin TW (2012) Light and vision in the deep-sea benthos: II. Vision in deep-sea crustaceans. J Exp Biol 215:3344–3353

Gómez-Tuena A, Orozco-Esquivel MT, Ferrari L (2007) Igneous petrogenesis of the Trans-Mexican Volcanic Belt. In: Alaniz-Álvarez SA, Nieto-Samaniego ÁF (eds) Geology of México: celebrating the centenary of the geological society of México: geological society of America special paper 422, pp 129–181

García-Machado E, Hernández D, García-Debrás A et al (2011) Molecular phylogeny and phylogeography of the Cuban cave-fishes of the genus *Lucifuga*: evidence for cryptic allopatric diversity. Mol Phylogenet Evol 61:470–483

Greenfield DW, Greenfield TA, Woods RL (1982) A new subspecies of cave-dwelling pimelodid catfish, *Rhamdia laticauda typhla* from Belize, Central America. Brenesia 19(20):563–576

Harms W (1921) Das rudimentäre Sehorgan eines Höhlendecapoden *Munidopsis polymorpha* Koelbel aus der Cueva de los Verdes auf der Insel Lanzarote. Zool Anz 52:101–114

Hendrickson DA, Krejca JK, Rodríguez Martinez JM (2001) Mexican blindcats genus *Prietella* (Siluriformes: Ictaluridae): an overview of recent explorations. Environ Biol Fishes 62:315–337

Hoenemann M, Neiber MT, Humphreys W et al (2013) Phylogenetic analysis and systematic revision of Remipedia (Nectiopoda) from Bayesian analysis of molecular data. J Crustac Biol 33(5):603–661. doi:10.1163/1937240X-00002179

Humphreys WF (2000) Relict faunas and their derivation. In: Wilkens H, Culver DC, Humphreys WF (eds) Ecosystems of the world: subterranean ecosystems, vol 30. Elsevier, Amsterdam, pp 417–432

Hunter RL, Webb MS, Iliffe TM et al (2007) Phylogeny and historical biogeography of the cave-adapted shrimp genus *Typhlatya* (Atyidae) in the Caribbean Sea and western Atlantic. J Biogeogr 35:65–75. doi:10.1111/j.1365-2699.2007.01767.x

Iliffe TM, Botoşăneanu L (2006) The remarkable diversity of subterranean Cirolanidae (Crustacea: Isopoda) in the peri-Caribbean and Mexican Realm. Bull Inst R Sci Nat Belg 76:5–26

Iliffe TM, Wilkens H, Parzefall J et al (1984) Marine lava cave fauna: composition, biogeography, and origins. Science 225:309–311

Iturralde-Vinent MA, García-Casco A, Rojas-Agramonte Y et al (2016) The geology of Cuba: a brief overview and synthesis. GSA Today 26(10):4–10. doi:10.1130/GSATG296A.1

Jaume D, Iliffe TM, Van der Ham JL (2013) A new *Psammogammarus* (Amphipoda: Eriopisidae) from anchialine pools on the Exuma Cays, Bahamas. Zootaxa 37(1):48–64

Joachim BL, Riesch R, Jeffery WR et al (2013) Pigment cell retention in cavernicolous populations of *Poecilia mexicana* (Poeciliidae). Bull Fish Biol 14:61–73

Juberthie-Jupeau L (1976) Sur le système neurosécréteur du pédoncule oculaire d'un décapode souterrain microphthalme *Typhlatya garciai* Chace. Ann Spéléol 31:107–114

Kindler P, Guillevic M, Baumgartner M et al (2014) Temperature reconstruction from 10 to 120 kyr b2k from the NGRIP ice core. Clim Past 10:887–902. doi:10.5194/cp-10-887-2014

Koenemann S, Iliffe TM (2013) Class Remipedia. In: von Vaupel Klein JC, Charmantier-Daures-M, Schram FR (eds) Treatise on zoology—anatomy, taxonomy, biology, The Crustacea, vol 4A. Leiden, Brill, pp 125–177

Kosswig C (1967) Über das Tempo evolutorischer Prozesse. Zoologische Beiträge 13:441–450

Mejía-Ortíz LM (2005) Adaptations to cave life in decapods from Oaxaca. Association for Mexican Cave Studies Bulletin 15, Austin, p 170

Mejía-Ortíz LM, Hartnoll RG, Viccon-Pale JA (2003) A new stygobitic crayfish from Mexico, *Procambarus cavernicola* (Decapoda: Cambaridae), with a review of cave-dwelling crayfishes in Mexico. J Crustac Biol 23(2):391–401

Mejía-Ortíz LM, Baldari F, López-Mejía M (2008) *Macrobrachium sbordonii* (Decapoda: Palaemonidae), A new stygobitic species of fresh water prawn from Chiapas, Mexico. Zootaxa 1814:49–57

Meyer-Rochow VB, Juberthie-Jupeau L (1987) An electron microscope study of the eye of the cave mysid *Heteromysoides cotti* from the island of Lanzarote (Canary Islands). Stygologia 3:24–34

Möller PR, Schwarzhans W, Iliffe TM et al (2006) Revision of the Bahamian cave-fishes of the genus *Lucifuga* (Ophidiiformes, Bythitidae), with description of a new species from islands on the Little Bahama Bank. Zootaxa 1223:23–46

Neiber MT, Hartke TR, Stemme T et al (2011) Global biodiversity and phylogenetic evaluation of Remipedia (Crustacea). PLoS One 6(5):e19627. doi:10.1371/journal.pone.0019627

Olesen J, Martinsen SV, Iliffe TM et al (2014) Remipedia. In: Martin J, Olesen J, Høeg J (eds) Atlas of crustacean larvae. John Hopkins University Press, Baltimore, pp 84–89

Palacios M, Voelker G, Rodriguez LA et al (2016) Phylogenetic analyses of the subgenus *Mollienesia* (Poecilia, Poeciliidae, Teleostei) reveal taxonomic inconsistencies, cryptic biodiversity, and spatio-temporal aspects of diversification in Middle America. Mol Phylogenet Evol 103:230–244

Parzefall J, Wilkens H (1972) Artbildung bei Höhlenfischen. Zoomorphology 73:63–79

Perdices A, Bermingham E, Montilla A et al (2002) Evolutionary history of the genus *Rhamdia* (Teleostei: Pimelodidae) in Central America. Mol Phylogenet Evol 25:172–189

Perdices A, Doadrio I, Bermingham E (2005) Evolutionary history of the synbranchid eels (Teleostei: Synbranchidae) in Central America and the Caribbean islands inferred from their molecular phylogeny. Mol Phylogenet Evol 37:460–473

Pérez-Moreno JL, Iliffe TM, Bracken-Grissom HD (2016) Life in the underworld: Anchialine cave biology in the era of speleogenomics. Int J Speleol 45(2):149–170

Peters N, Peters G, Parzefall J et al (1973) Über degenerative und konstruktive Merkmale bei einer phylogenetisch jungen Höhlenform von *Poecila sphenops*. Int Rev Gesamten Hydrobiol 58:417–436

Porter ML, Dittmar K, Pérez-Losada M (2007) How long does evolution of the troglomorphic form take? Estimating divergence times in *Astyanax mexicanus*. Acta Carsol 36(1):173–182

Pregill GK, Olson SL (1981) Zoogeography of West Indian vertebrates in relation to Pleistocene climatic cycles. Annu Rev Ecol Syst 12:75–98

Rasmussen TL, Thomsen E, Moro M (2016) North Atlantic warming during Dansgaard-Oeschger events synchronous with Antarctic warming and out-of-phase with Greenland climate. Sci Rep 6:1–12. doi:10.1038/srep20535

Schemmel C (1977a) Zur Morphologie und Funktion der Sinnesorgane von *Typhliasina pearsei* (Hubbs) (Ophidioidea, Teleostei). Zoomorphology 87:191–202

Stanley SM, Campbell D (1981) Neogene mass extinction of Western Atlantic molluscs. Nature 293:457–459

Stegner MEJ, Stemme T, Iliffe TM et al (2015) The brain in three crustaceans from cavernous darkness. BMC Neurosci 16:1–28. doi:10.1186/s12868-015-0138-6

Weber A (2003) Taxonomie und Artstatus höhlenbewohnender und oberirdischer Welse der Gattung *Rhamdia* (Pimelodidae, Teleostei). Dissertation, University of Hamburg

Weber A, Wilkens H (1998) *Rhamdia macuspanensis*: a new species of troglobitic pimelodid catfish (Siluriformes, Pimelodidae) from a cave in Tabasco, Mexico. Copeia:998–1004

Weber A, Allegrucci G, Sbordoni V (2003) *Rhamdia laluchensis*, a new species of troglobitic catfish (Siluriformes, Pimelodidae) from Chiapas, Mexico. Ichthyol Explor Freshwaters 14:273–280

Weyl R (1964) Die paläogeographische Entwicklung des Mittelamerikanisch-Westindischen Raumes. Geol Rundsch 54:1213–1240

Wilhelm O, Ewing M (1972) Geology and history of the Gulf of Mexico. Geol Soc Am Bull 83:575–600

Wilkens H (1972) Zur phylogenetischen Rückbildung des Auges Cavernicoler: Untersuchungen an *Anoptichthys jordani* (=*Astyanax mexicanus*) Characidae, Pisces. Ann Spéliol 27:411-432

Wilkens H (1973) Über das phylogenetische Alter von Höhlentieren. Untersuchungen über die cavernicole Süßwasserfauna Yucatans. Z Zool Syst Evol 11:49–60

Wilkens H (1979) Reduktionsgrad und phylogenetisches Alter: Ein Beitrag zur Besiedlungsgeschichte der Limnofauna Yukatans. Z Zool Syst Evol 17:262–272

Wilkens H (1982) Regressive evolution and phylogenetic age: the history of colonization of freshwaters of Yucatan by fish and Crustacea. Texas Mem Mus Bull 28:237–243

Wilkens H (1986) The tempo of regressive evolution: studies of the eye reduction in stygobiont fishes and decapod crustaceans of the Gulf Coast and Western Atlantic region. Stygologia 2:139–143

Wilkens H (2001) Convergent adaptations to cave life in the *Rhamdia laticauda* catfish group (Pimelodidae, Teleostei). Environ Biol Fishes 62:251–261

Wilkens H, Strecker U, Yager J (1989) Eye reduction and phylogenetic age in ophidiiform cave fish. Z Zool Syst Evol 27:126–134

Wilkens H, Junge P, Langecker TG (1991) Speciation of troglobites: studies in the San Antonio cave (Oaxaca, Mexico). Int J Speleol 20:1–13

Wilkens H, Iliffe TM, Oromí P et al (2009) The Corona lava tube, Lanzarote: geology, habitat diversity and biogeography. Mar Biodivers 39:155–167. doi:10.1007/s12526-009-0019-2

Wilkens H, Parzefall J (1974) Die Ökologie der Jameos del Agua (Lanzarote). Zur Entwicklung limnischer Höhlentiere aus marinen Vorfahren. Ann Spéléol 29:419–434

Yager J (1981) Remipedia, a new class of Crustacea from a marine cave in the Bahamas. J Crustac Biol 1(3):328–333

# Surface and Cave Populations of Mexican *Astyanax* 4


**Abstract**

The Neotropic large-eyed and well-pigmented diurnal characid fish *Astyanax* has developed a series of cave populations in Northeastern Mexico. These divide morphologically into a group of strongly eye- and pigment-reduced (SEP) cave populations and another one characterized by variable eye size and pigmentation (VEP cave populations). Molecular and biogeographic data imply that they derive from the Neotropic *Astyanax* surface fish, which were able to invade North America up to the Rio Grande drainage after the closure of the Central American land bridge. Its recent distributional pattern is strongly influenced by Pleistocene climatic changes and is characterized by regional extinction and recolonization from the warmer south and/or survival in climatically buffered refuges. An example of this are the SEP cave fish populations, which according to cytochrome b analysis do not cluster with the surface fish from neighboring rivers and creeks but with fish from a remote location about 500 km away in the Central Mexican Plateau. In line with this, they do not group with either the VEP cave fish or with surface fish from the cave area, and based on microsatellites and SNP studies, they exhibit relation to populations from southern Mexico and Belize. The SEP cave fish and some relic surface fish populations from isolated locations all over Mexico derive from the oldest invasion. In contrast, based on cytochrome b studies, the VEP cave populations cluster with the recent surface fish from the cave area, which is widespread in Northern Mexico. The VEP cave populations derive from a more recent invasion of surface fish into Northern Mexico. In particular, the differing degree of eye reduction between SEP and VEP cave fish reflects the different times of cave entry. Cave colonization in VEP and SEP cave populations took place in parallel and resulted in multiple convergent evolutions.





H. Wilkens, U. Strecker, *Evolution in the Dark*, DOI 10.1007/978-3-662-54512-6_4

## 4.1 *Astyanax* Surface Fish

The characid *Astyanax* is a Neotropic carnivorous midwater fish that possesses large eyes and well-developed pigmentation. It is widespread in all types of freshwater habitats in the coastal lowlands but has also penetrated several streams in the highlands of Mexico as far north as the Rio Grande in Texas and its tributaries (Miller and Smith 1986). Its taxonomy is still under dispute. Based on meristic and morphometric differences as well as on molecular genetic analyses, varying numbers of species have been suggested (Eigenmann 1917; Géry 1977; Reis et al. 2003; Schmitter-Soto 2016), which often do not meet the rules of the International Code of Zoological Nomenclature (e.g. Ornelas-García et al. 2008) (Fig. 4.1) (see Sect. 5.5).

## 4.2 *Astyanax* Cave Fish

In three separate geographically adjacent limestone ridges located at the eastern margin of the Mexican Sierra Madre Oriental del Norte, the Sierra de El Abra (El Abra populations), Sierra de la Colmena (Rio Subterráneo population), and Sierra de Guatemala (Guatemala populations), a series of cave populations have originated (Table 4.1, Figs. 4.2 and 4.3). Only one *Astyanax* cave population occurs far away from this area, in Central Mexico (Espinasa et al. 2001). Since the discovery of the first blind cave characid *Anoptichthys jordani* (Hubbs and Innes 1936), the Chica fish, about 30 caves inhabited by *Astyanax* cave fish have been detected and explored to date (Mitchell et al. 1977).

The largest number of populations is found in the Sierra de El Abra (Elliott 2015a, b; Gross 2012; Mitchell et al. 1977) (Fig. 4.3). Their most obvious features are external eyelessness caused by the strongly reduced eye rudiments deeply sunken into the eye orbit as well as paleness resulting from decreased numbers of melanophore colour cells and reduced melanin content (Table 4.1) (Mitchell et al.

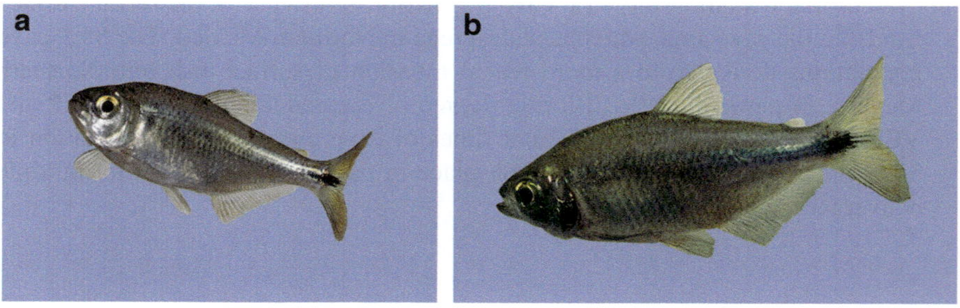


**Fig. 4.1** *Astyanax* surface fish from Rio Teapao (Southern Mexico) (cf. "*A. aeneus*") and Cenote Dzibilchaltún (Yucatán) (cf. "*A. altior*")

**Table 4.1** Distribution and characteristics of the main *Astyanax* cave populations

| | | Phylogenetically | | Introgressed by | | Eyes | | Melanophores | | | |
|---|---|---|---|---|---|---|---|---|---|---|---|
| Cave population | Location | Old/ SEP | Young/ VEP | SEP | VEP | Strongly reduced | Variably reduced | Number reduced | Brown gene | Albino gene | Surface fish in cave |
| Pachón | Sierra de El Abra | X | | | X | X | | X | X | X | |
| Yerbaniz | Sierra de El Abra | X | | | X | X | | X | X | X | Many |
| Sabinos | Sierra de El Abra | X | | | | X | | X | X | | |
| Arroyo | Sierra de El Abra | X | | | | X | | X | X | | |
| Tinaja | Sierra de El Abra | X | | | | X | | X | X | | |
| Pichijumo | Sierra de El Abra | X | | | X | X | | X | X | | Few |
| Piedras | Sierra de El Abra | X | | | | X | | X | X | | |
| Curva | Sierra de El Abra | X | | | | X | | X | X | | |
| Toro | Sierra de El Abra | X | | | | X | | X | X | | |

(continued)

**Table 4.1** (continued)

| Cave population | Location | Phylogenetically | | Introgressed by | | Eyes | | Melanophores | | | Surface fish in cave |
|---|---|---|---|---|---|---|---|---|---|---|---|
| | | Old/ SEP | Young/ VEP | SEP | VEP | Strongly reduced | Variably reduced | Number reduced | Brown gene | Albino gene | |
| Chica | Sierra de El Abra | | X | X | | | X | ? | ? | | ? (Few or none) |
| Cuates | Sierra de El Abra | | X | X ? | | | X | ? | ? | | |
| Micos | Sierra Colmena | | X | | | | X | | | X[b] | Many |
| Caballo | Sierra Guatemala | | X | | | | X | ? | | | |
| Molino | Sierra Guatemala | | X | | | | X[a] | X | | X | |

[a]Only genetically variable
[b]Not manifested
*SEP* strongly eye- and pigment-reduced, *VEP* variable eye size and pigmentation, *?* uncertain

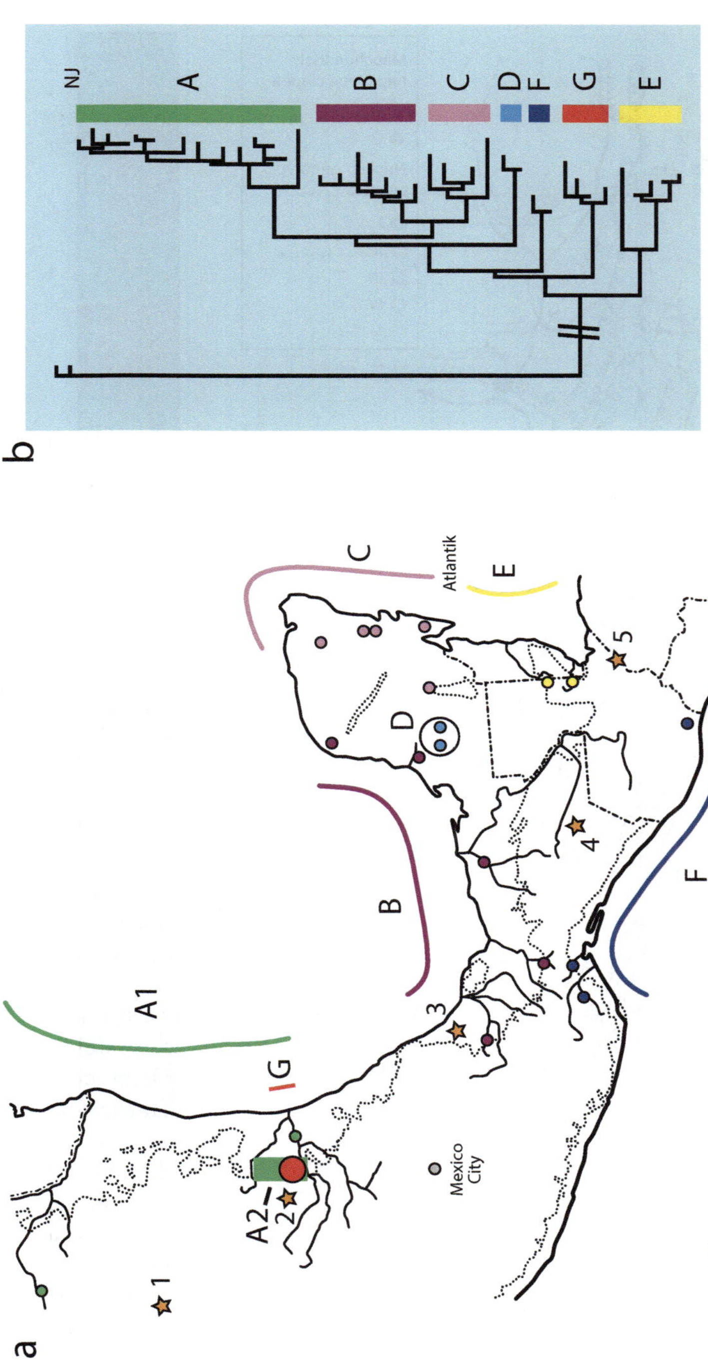


**Fig. 4.2** Pattern of geographical distribution of surface and cave *Astyanax* mtDNA lineages occurring in Mexico, Guatemala, and Belize (**a**) and Bayesian consensus tree (**b**) illustrating the phylogenetic relationships among 53 cytochrome b mtDNA haplotypes found in 35 Astyanax surface and cave populations. The clades A to F include surface populations. The clade from the northern Gulf Coast (A) is split into one subclade (A1) containing all surface populations from the Rio Panuco drainage and one haplotype found in a population from the Rio Grande system, and the subclade A2 including three of the strongly reduced SEP cave populations Pachón, Pichijumo, and Yerbaniz, as well as the VEP cave populations Molino, Chica, and Micos. Clade G includes the SEP Piedras, Tinaja, Sabinos, and Curva cave populations. 1–5 = Surface populations from Ríos Aguanaval/Mezquital (1), Rascón/Tamasopo (2), Río Máquinas (3), Lagunas de Montebello (4), Ríos Polochic, Grijalva, Usumacinta (5). *Dotted line* indicates 200 m contour line, *SEP* strongly eye- and pigment-reduced, *VEP* variably eye- and pigment-reduced (adapted from Strecker et al. 2004; Ornelas-García et al. 2008)

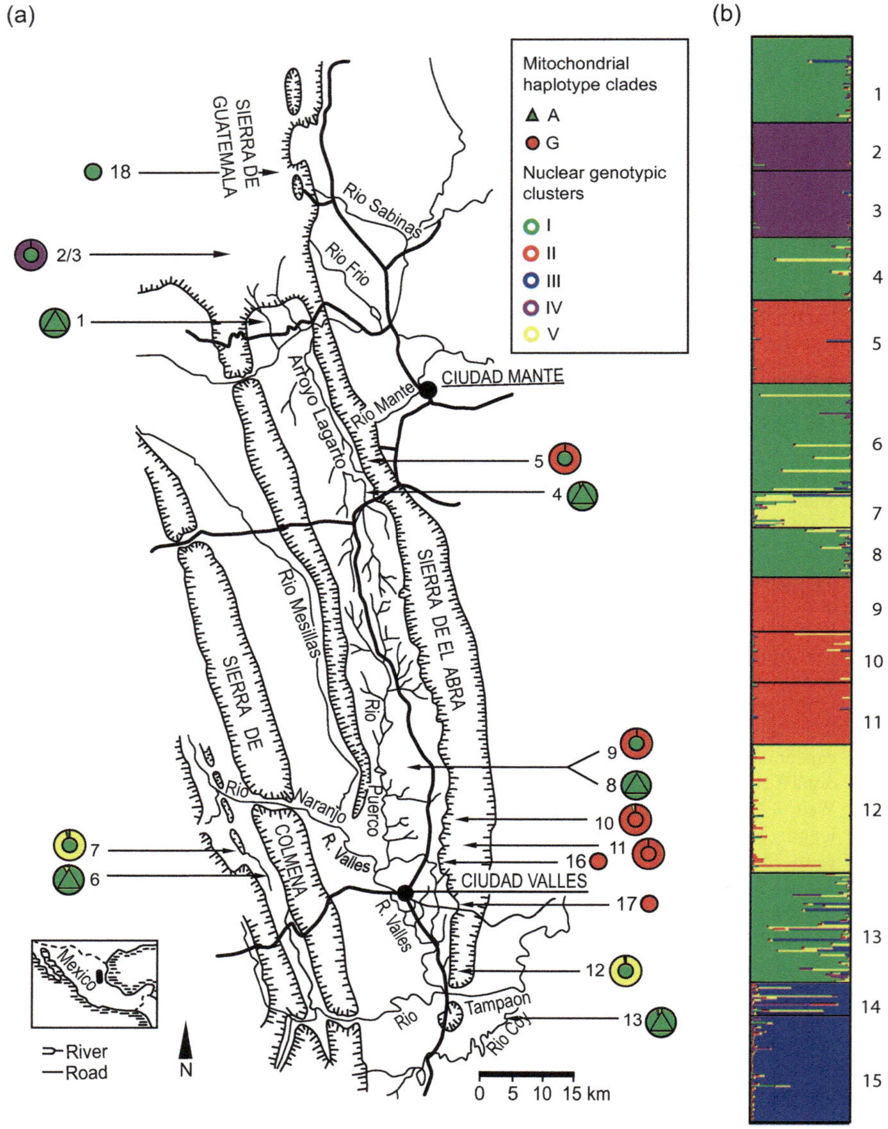


**Fig. 4.3** Geographic distribution and genetic variation in Mexican surface and cave *Astyanax* populations. (**a**) Map of sampling sites in the cave region in Northeastern Mexico. *Outer circles* represent nuclear genotypic clusters delimited using STRUCTURE for K = 5; *inner circles* and *triangles* represent haplotype lineages of cave and surface populations, respectively. (**b**) Results of the admixture analysis with STRUCTURE for K = 5. Numbers refer to populations: 1. Rio Coahuila, 2. Caballo Moro cave fish with sunken eyes, 3. Caballo Moro cave fish with eyes, 4. Pachón surface, 5. Pachón cave, 6. Micos surface, 7. Micos cave, 8. Yerbaniz cave (surface fish washed into the cave), 9. Yerbaniz cave (cave fish), 10. Sabinos cave, 11. Tinaja cave, 12. Chica cave, 13. Rio Coy, 14. Rio Coatzacoalcos drainage, 15. Mahajual (Eastern Yucatán), 16. Piedras cave, 17. Curva cave, 18. Molino cave (adapted from Strecker et al. 2012)

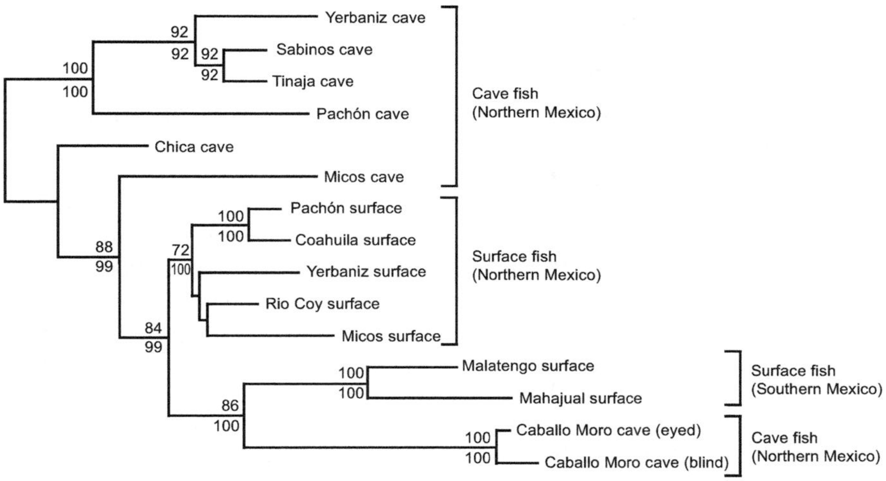


**Fig. 4.4** Neighbour joining tree based on Nei's $D_A$ distance from six microsatellite loci of *Astyanax* cave and surface fish from Northeastern Mexico and surface fish from Southern Mexico. Values along branches indicate bootstrap values either based on $D_A$ distances (*above*) or $F_{ST}$ values (*below*) (adapted from Strecker et al. 2012)

1977; Wilkens 1988, 2007). They were therefore denominated strongly eye- and pigment-reduced (SEP) populations (Strecker et al. 2004, 2012; Wilkens 1988) and based on cytochrome b studies classified as phylogenetically old. According to microsatellite analysis, the El Abra populations build a cluster of their own (Bradic et al. 2012; Strecker et al. 2012) (Fig. 4.4).

A second group exhibits variability in the size of their eyes (which, except for the Molino cave fish, are not or just slightly overgrown by body tissue and are still visible externally) and the dark melanin pigmentation. Because of their variability of eyes and pigmentation they are called VEP populations (Strecker et al. 2004). Except for the Chica and probably also the geographically adjacent Los Cuates populations, all others are found outside the Sierra de El Abra in the S. de Colmena (Micos or Rio Subterraneo population) and the S. de Guatemala (e.g. Caballo Moro and Molino population) (Table 4.1). All were originally designated 'mixed fish populations' by Mitchell et al. (1977), because it was believed that these populations were composed of surface fish, cave fish, and their hybrids. In contrast to this, the variability of eyes and pigmentation was alternatively interpreted as resulting from their phylogenetically young age (Table 4.1) (Kosswig 1965; Strecker et al. 2003, 2004, 2012; Wilkens 1988).

## 4.3 Phylogeography and Speciation of Surface and Cave *Astyanax*

### 4.3.1 Invasion of Surface *Astyanax* from South America and Origin of the Cave Forms

It is assumed that surface *Astyanax* as a Neotropic primary freshwater fish could not invade North America before the closure of the Middle American land bridge in the late Pliocene (Bussing 1985; Bermingham and Martin 1998; Myers 1966; Perdices et al. 2002; Reeves and Bermingham 2006). Phylogeographic studies based on mtDNA revealed several distinctive haplotype lineages in *Astyanax* surface fish in Northern Central America, which mainly represent geographical patterns of distribution (Strecker et al. 2004; Ornelas-García et al. 2008) (Fig. 4.2). Some cover a broad geographic range like lineages A and B, north and south of the Trans-Mexican Volcanic Belt (TMVB), respectively, whereas others are locally restricted as, for example, lineage D on Yucatán or Montebello (4 in Fig. 4.2). Interestingly, in Northeastern Mexico the SEP cave populations (clade G) do not cluster with their neighbouring surface populations (clade A2), nor do the Belizean populations (clade E) group with their geographic neighbours from the Yucatán (clades B, C and D) or with *Astyanax* from locations 4 and 5 (Strecker et al. 2004; Ornelas-García et al. 2008) (Fig. 4.2).

The current distribution pattern of haplotype lineages can be explained by several invasions of surface *Astyanax* from the South. This was probably caused by Pleistocene climate changes and their concomitants like fluctuating temperatures and precipitation leading to large-scale extinction and distributional fragmentation. During warmer periods, recolonization of different genetic lineages took place either from the South or from isolated refugia within the former distributional area (Hausdorf et al. 2011; Strecker et al. 2004). An example of such glacial refuge may be provided by a warm spring in Northwestern Mexico, which to date, at 2000 m altitude, still represents the highest elevation attained by *Astyanax* surface fish in Mexico north of the TMVB (Miller and Smith 1986). It lies between the endorheic drainage of the Rios Aguanaval or Rio Nazas and the Pacific affluent Rio Mezquital (Fig. 4.2), which contain surface *Astyanax* carrying, beside the wide-spread haplogroup A, the same haplotype lineage G as the SEP cave fish in caves about 500 km away by air. Similarly, the SEP cave fish (haplotype lineage G) could survive in caves in the Sierra de El Abra because the cave habitat in general is characterized by constancy of temperature enabling species to overcome a temporarily fluctuating climate (Juberthie 2000). This environmental characteristic is proven by the co-occurrence of a series of crustacean cave species like palaemonid shrimps (genus *Troglomexicanus*), cirolanid isopodes (genus *Speocirolana*) (Fig. 3.6), or mysids (genus *Speleomysis*). These species are much older inhabitants in these caves and had already colonized the Pliocene anchialine caves situated along the Sierra de El Abra before the arrival of *Astyanax* surface fish in the cave region (see Sect. 3.2). The SEP cave haplotype lineage G is a sister group of the before-mentioned geographically remote surface populations from Aquanaval and

Mezquital (Fig. 4.2). They are more closely related than either are to two other relic surface populations, Rascón and Tamasopo, found close to the cave area by Ornelas-García et al. (2008).

Additionally, three other lineages occur far south of the TMVB in the drainage systems Polochic-Grijalva-Usumacinta, Lagunas de Montebello, and Río Máquinas (Ornelas-García et al. 2008). These four lineages build a polytomy suggesting that the ancestral lineage spread rapidly across Northern America, as was similarly shown for characiform fish in Mesoamerica (Reeves and Bermingham 2006). The large genetic distance of 12–17 mutations between these lineages indicates a rather old separation. The four haplotype lineages form a basal clade (Fig. 4.2) (Ornelas-García et al. 2008), suggesting that they descended from the first invasion from the South to north of the TMVB (Strecker et al. 2004). This is supported by the high level of troglomorphism found in all SEP populations, given that the degree of troglomorphism and phylogenetic age are correlated (see Chap. 3) (Culver and Wilkens 2000).

The haplotype lineage A is widely spread north of the TMVB and represents a more recent invasion. It includes surface populations as well as SEP and VEP cave fish. As will be shown below, it is suggested that the SEP populations, which are grouping in this lineage, are the result of introgressive hybridization (see Sect. 5.4).

Notably, SEP cave populations like Piedras, Sabinos, Tinaja, and Curva exclusively occur in the Sierra de El Abra and are missing outside it in the geographically separated Sierra de la Colmena and Sierra de Guatemala, where only VEP cave populations and none of the Pliocene crustacean cave genera typical of the Sierra de El Abra have been found as yet. It is hypothesized that this biogeographical difference might be explained by the microclimatic and/or hydrological differences of these Sierras. It is worthwhile considering whether the SEP *Astyanax* cave populations that potentially originally existed in the other Sierras became extinct because of the inflow of cool water from higher altitudes during glacial advances, as is proposed for the cave fauna located south of the TMVB (see Sect. 3.3).

In contrast with the Sierra de la Colmena and de Guatemala, no VEP *Astyanax* cave populations seem to exist in the Sierra de El Abra. However, this is only at first glance. It is proposed that VEP cave populations have also arisen in the Sierra de El Abra, because, in karst, new caves are continuously being formed and can be colonized. Such cave populations may have introgressed and merged with already existing SEP cave populations due to the erosional dynamics of karst. Proof for this is provided by the SEP Pachón and Yerbaniz cave populations, which cluster with the recent surface fish because of mitochondrial capture, but based on their nuclear genotype belong to SEP cave fish (see Chap. 5). As the only exception, Chica can be characterized as a VEP cave fish in the Sierra de El Abra, although in this population too, introgression from an unknown SEP cave population was detected by microsatellite analysis (Strecker et al. 2003) (Fig. 4.3).

The date of cave entry of the *Astyanax* cave fish is still unknown and controversial. Based on allozyme variability, the SEP cave populations were calculated to be very young, having a post-Pleistocene age (Avise and Selander 1972). In contrast, gene differentiation calculations between populations of unequal sizes revealed an approximate divergence time between 710,000 ± 460,000 years ago for the Pachón and 525,000 ± 330,000 years ago for the Sabinos cave fish (Chakraborthy and Nei 1974).

According to the calibration rate of 1.5% (Zardoya and Doadrio 1999) the cytochrome b analyses of *Astyanax* revealed a divergence time of between 1.8 and 4.5 million years ago (mya) with a mean of 3.1 mya (Strecker et al. 2004), which is a date concurring with that of the final closure of the Middle American land bridge at the end of the Pliocene (Reeves and Bermingham 2006; Picq et al. 2014). This date is also confirmed by the second Neotropic salt intolerant primary freshwater fish genus *Rhamdia* to reach North America, including the *R. laticauda* clade, which is sister species to a series of cave species (Wilkens 2001), followed by a rapid expansion in Central America between 2.9 and 2.5 mya (Perdices et al. 2002).

In contrast, Ornelas-García et al. (2008, 2014) assume an earlier invasion of surface *Astyanax* during the Miocene at about 7.8–8.1 mya using a different calibration rate. However, biogeographical data from the totally karstic Yucatán peninsula, a landscape characterized by the complete absence of surface rivers, support the more recent invasion time as calculated by Strecker et al. (2004). The geographical distribution of the salt-intolerant primary freshwater fish *Astyanax* is exclusively restricted to freshwater habitats situated in the narrow Pleistocene coastal plain (Fig. 4.5). *Astyanax* is not found in the cenotes of the interior of northern Yucatán, where neotropic salt-tolerant secondary freshwater fish species like *Cichlasoma urophthalmus* (Cichlidae) and the live-bearing toothcarp *Gambusia puncticulata* (Poeciliidae) are common, however (Wilkens 1979). This diverging distribution pattern of primary and secondary fish species results from the fact that the salt-tolerant secondary freshwater species had been able to invade North America long before the final uplift of the Middle American land bridge (Myers 1966; Martin and Bermingham 2000). They could already colonize the interior of the present Yucatán peninsula while, during Mio- and Pliocene, the sea slowly retreated from the limestone platform in a northeasterly direction. The exclusive existence of *Astyanax* in the coastal zones demonstrates that this fish was not yet present in North America, while the secondary freshwater fish already were. As an exception, the hepapterid catfish *Rhamdia guatemalensis*, another Neotropic salt-intolerant primary freshwater fish that arrived at the same time as *Astyanax* in North America (Perdices et al. 2002), was able to later reach the cenotes of the interior of Northern Yucatán. In contrast to *Astyanax*, this fish is a nocturnal and troglophile species. It took advantage of the vast aquatic underground pathways characteristic of the karstic Yucatán peninsula for dispersal (Fig. 4.5) (Wilhelm and Ewing 1972; Wilkens 1982, 1986).

Summarizing, it is suggested that *Astyanax* was able to invade North America only after the closure of the land bridge about 3 mya, as is proposed for other tropical primary freshwater fishes (Bermingham and Martin 1998; Reeves and Bermingham 2006; Perdices et al. 2002), whereas salt-tolerant, so-called secondary freshwater fish like cichlids or poecilids could have already dispersed to the North about 8 mya (Martin and Bermingham 2000; Myers 1966).

Thus, based on the assumption that the Neotropic surface *Astyanax* only reached Northern Mexico after the final uplift of the Central American land bridge, cave colonization and origin of the SEP cave forms in Northeastern Mexico can be dated

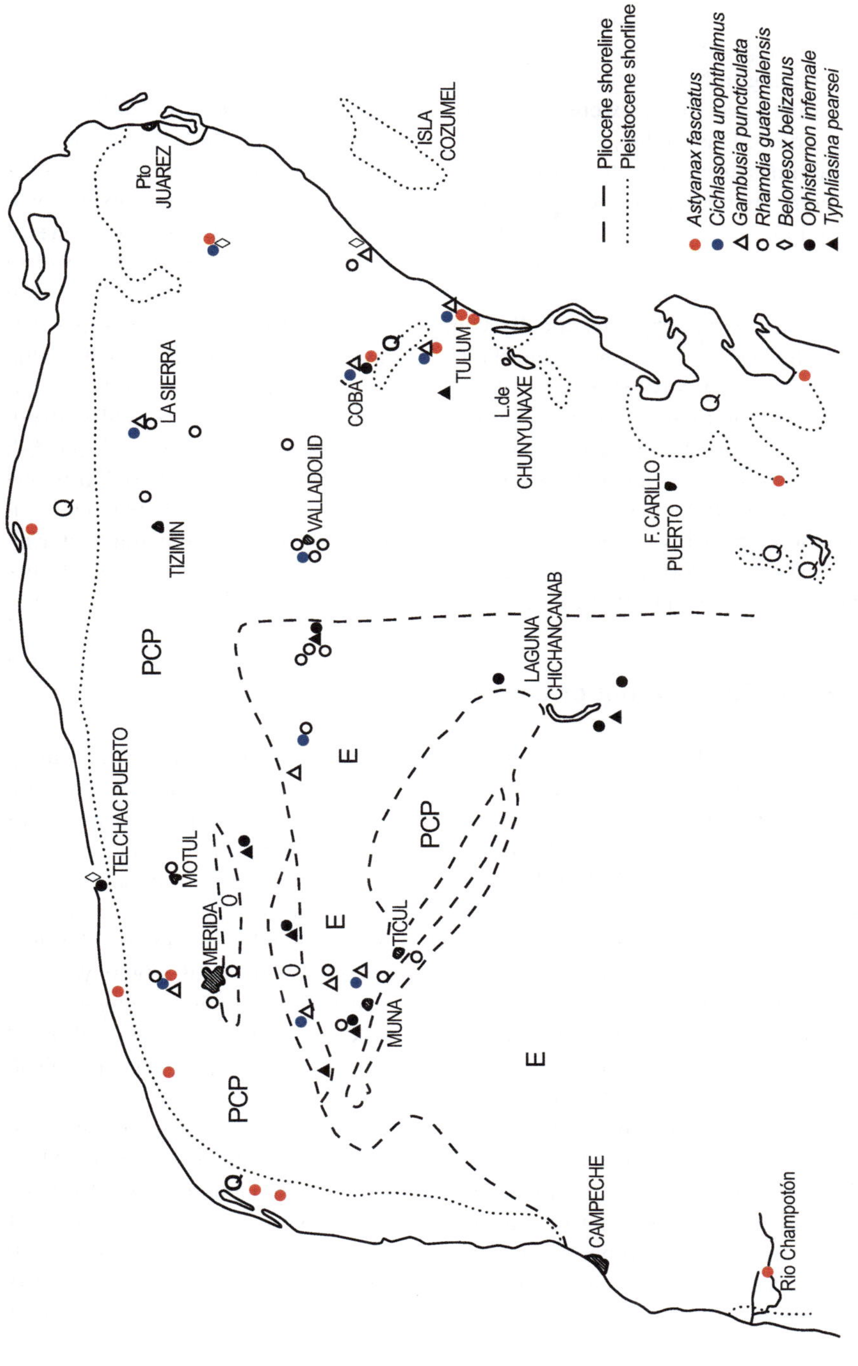


**Fig. 4.5** Distribution of the main freshwater fish species in Yucatán. *Q* Pleistocene, *PCP* Mio/Pliocene, *O* Oligocene and *E* Eocene sediments (adapted from Wilkens 1979; Wilhelm and Ewing 1972)

to have taken place around this time, but no earlier. This possibility is corroborated by the co-occurrence with Pliocene marine cave relics, which indicates that surface *Astyanax* in principle could have colonized the caves in the Sierra de El Abra soon after arriving in North America about 3.0 mya. It cannot be ruled out, though, that the marine cave species were better cold adapted than *Astyanax*, which has Neotropic origin. Due to this, cave populations of this fish might have become extinct several times during Pleistocene climatic coolings. In addition, these calculations of cave entry are uncertain, because it can generally be assumed that the cytochrome b calibration data only reveal the splitting time between surface lineages, one of which became the ancestor of the cave forms at an unknown later date (Trontelj et al. 2007).

As concerns the phylogenetically younger *Astyanax* VEP cave populations, the split of lineages for the Micos and the Chica cave fish was calculated as 0.26 and 0.39 mya, respectively (Strecker et al. 2004). However, when comparing the similar intermediate and variable degrees of eye regression of the Micos, Caballo Moro, and Chica cave fish populations with those of the group of lesser eye-reduced fish and crustacea occurring in Mexico south of the TMVB and the Peri-Caribbean islands of the Bahamas (see Chap. 3), it can also be concluded that the VEP populations possibly only originated after the last glacial maximum during Wisconsin at 27,000 to 24,000 years ago, or even only after the Younger Dryas period at 12,900 to 11,700 years ago.

### 4.3.2 Multiple Origin of Cave Forms

Caves as isolated habitats are usually colonized in parallel by the same widely spread ancestral surface species. As a result, convergent adaptation to cave life starting at the same or even subsequent times may arise. Examples for this have been demonstrated for the freshwater shrimp genus *Macrobrachium* occurring in Mexico south of the TMVB or the crayfish genus *Procambarus* from south of the TMVB and Florida. Also, the hepapterid surface catfish *Rhamdia laticauda* occurring south of the TMVB is sister species to a still rising number of cave species, five of which have as yet been taxonomically described (Weber et al. 2003; Wilkens 2001). In Northeastern Mexico, the ictalurid cave catfish *Prietella phreatophila* and *P. lundbergi*, which both derive from the same surface ancestor, have colonized caves distributed over two geographically separate limestone areas (Hendrickson et al. 2001; Wilcox et al. 2004).

The *Astyanax* cave populations, too, are characterized by this evolutionary pattern. The caves in the three different limestone ridges, the Sierra de El Abra, S. de Colmena, and S. de Guatemala, are geographically separated from each other, which indicates independent colonization. This was by at least two invasions of surface fish at different times from the South of Mexico, resulting in SEP and VEP cave populations. For the Sierra de El Abra, it has long been debated whether the cave populations occurring in this karst area are the result of secondary dispersal of an already cave-adapted ancestor or if they have multiple origins (Avise and Selander 1972; Espinasa and Borowsky 2001; Dowling et al. 2002; Mitchell et al.

1977; Strecker et al. 2003). Based on allozyme studies, it was concluded for the cave fish of the Sierra de El Abra that "the eyeless and unpigmented condition is believed to have evolved in whole or part prior to the present-day subdivision of the populations" (Avise and Selander 1972). In contrast, Mitchell et al. (1977) suggested, on the basis of hydrological data from the Sierra de El Abra, that a multiple colonization hypothesis is more plausible. This was corroborated by molecular studies. It is assumed that the *Astyanax* surface fish invaded different caves separately and were submitted to convergent evolution (Bradic et al. 2012; Coghill et al. 2014; Strecker et al. 2004, 2012).

Confirmation of convergent evolution of the different SEP cave populations was for the first time shown by the development of larger and better differentiated eyes in the F1 crossing between the SEP Pachón and Sabinos cave fish revealing differences of the genetic basis of eye reduction (Wilkens 1971). Similarly, differences in the respective size ranges of the F2 crossings of the SEP Pachón, Yerbaniz, Piedras, and Curva occurring in the Sierra de El Abra with the VEP Molino cave fish from the Sierra de Guatemala demonstrate that even between the four SEP populations found in relatively close proximity in the Sierra de El Abra, eye genes are submitted to different regressive mutations, to a certain extent (Wilkens and Strecker 2003). The same was described for the pigment *albino* (*Oca2*) and *brown* genes (*Mc1r*) (Protas et al. 2007; Gross et al. 2009; Gross and Wilkens 2013; Stahl and Gross 2015, see Sects. 6.20.2 and 6.20.3). Nonetheless, the fact that introgressive hybridization with mitochondrial capture has taken place makes it probable that in the Sierra de El Abra, due to the dynamic geohydrological development of its karst system, cave populations were submitted to underground migration (Espinasa and Espinasa 2015) and some of them have merged into a single one.

### 4.3.3 Population Genetic Diversity of Cave Populations

With rare exceptions, as, for example, found in the troglobitic bivalve *Congeria kusceri* (Dreissenidae) (Bilandžija et al. 2013; Stepien et al. 2001), cave species are characterized by low genetic variability (Caccone et al. 2000; Culver 1982; Culver et al. 1995; Culver and Pipan 2009; Kane et al. 1992). This is also valid for the *Astyanax* cave fish in comparison with the surface fish. Low mtDNA variability found in the SEP Curva, Pachón, Yerbaniz, Tinaja, and Sabinos cave fish and in the VEP Molino and Micos cave populations is congruent with previous data obtained from allozymes in the Pachón and Sabinos cave fish (Avise and Selander 1972; Dowling et al. 2002; Peters et al. 1975; Strecker et al. 2004). Comparable results are provided by microsatellite and transcriptome studies (Bradic et al. 2012; Hinaux et al. 2013; Strecker et al. 2003, 2012). Furthermore, intrapopulational nucleotide difference of opsin genes is the highest in the surface fish and decreases in Micos and Pachón cave fish, in that order (Yokoyama et al. 1995). It is assumed that the low genetic variability of the *Astyanax* cave populations results from repeated population bottlenecks resulting from temporary food scarcity. This is caused by the *Astyanax* caves receiving food input only once a year during the rainy season. At that time, specimens of surface *Astyanax* as well as of other surface species and plant material are washed into the cave systems. During the rest of the year the creeks at the surface are dry and no water flows from them into the underground caves. At that time, food scarcity arises as is obviously exemplified by the presence of starving *Astyanax* surface fish (see Fig. 5.3). Further proof of periodic bottlenecking in the *Astyanax* caves is provided by the change from polygenic in the surface to monogenic sex determination in the cave fish (see Sect. 6.3) As the only exception, the Chica cave receives rich food input from large bat roosts, which is so ample that even a large number of crayfish can coexist.

## References


Avise JC, Selander RK (1972) Genetics of cave-dwelling fishes of the genus *Astyanax*. Evolution 26:1–19

Bermingham E, Martin AP (1998) Comparative mtDNA phylogeography of neotropical freshwater fishes: testing shared history to infer the evolutionary landscape of lower Central America. Mol Ecol 7:499

Bilandžija H, Parkhurst A, Jeffery WR (2013) A potential benefit of albinism in *Astyanax* cavefish: downregulation of the *oca2* gene increases tyrosine and catecholamine levels as an alternative to melanin synthesis. PLoS One 8:1–14

Bradic M, Beerli P, García-de León F et al (2012) Gene flow and population structure in the Mexican blind cavefish complex (*Astyanax mexicanus*). BMC Evol Biol 12:1–16

Bussing WA (1985) Patterns of distribution of the Central American Ichthyofauna. In: Stehli FG, Webb SD (eds) The great American biotic interchange, vol 4. Plenum Press, New York, pp 453–473

Caccone A, Allegrucci G, Cesaroni D et al (2000) Genetic variability and divergence between cave dwelling populations of *Typhlocirolana* (Crustacea, Isopoda) from Majorca and Sicily. Biochem Syst Ecol 14:215–221

Chakraborthy R, Nei M (1974) Dynamics of gene differentiation between incompletely isolated populations of unequal sizes. Theor Popul Biol 5:460–469

Coghill LM, Hulsey CD, Chaves-Campos J et al (2014) Next generation phylogeography of cave and surface *Astyanax mexicanus*. Mol Phylogenet Evol 79:368–374

Culver CD (1982) Cave life. Harvard University Press, Cambridge

Culver CD, Pipan T (2009) The biology of caves and other subterranean habitats. Oxford University Press, New York

Culver DC, Wilkens H (2000) Critical review of the relevant theories of the evolution of subterranean animals. In: Wilkens H, Culver DC, Humphreys WF (eds) Ecosystems of the world: subterranean ecosystems, vol 30. Elsevier, Amsterdam, pp 381–398

Culver CD, Kane TC, Fong DW (1995) Adaptation and natural selection in caves. Harvard University Press, Cambridge

Dowling TE, Martasian DP, Jeffery WR (2002) Evidence for multiple genetic forms with similar eyeless phenotypes in the blind cavefish, *Astyanax mexicanus*. Mol Biol Evol 19:446–455

Eigenmann CH (1917) The American Characidae (Part I). Mem Mus Comp Zool 43:1–102

Elliott RE (2015a) Cave biodiversity and ecology of the Sierra de El Abra region. In: Keene AC, Yoshizawa M, McGaugh SE (eds) Biology and evolution of the Mexican cavefish. Elsevier, Amsterdam, pp 59–75

Elliott RE (2015b) Cave exploration and mapping in the Sierra de El Abra region. In: Keene AC, Yoshizawa M, McGaugh SE (eds) Biology and evolution of the Mexican cavefish. Elsevier, Amsterdam, pp 9–39

Espinasa L, Borowsky RL (2001) Origins and relationship of the blind Mexican tetra, *Astyanax fasciatus*, in the Sierra de El Abra. Environ Biol Fishes 62:233–237

Espinasa l, Espinasa M (2015) Hydrogeology of caves in the Sierra de El Abra Region. In: Keene AC, Yoshizawa M, McGaugh SE (eds) Biology and evolution of the Mexican cavefish. Elsevier, Amsterdam, pp 41–57

Espinasa L, Rivas-Manzano P, Pérez HE (2001) A new blind cave fish population of genus *Astyanax*: geography, morphology and behaviour. Environ Biol Fishes 62:339–344

Géry J (1977) Characoids of the world. TFH Publications, Neptune City

Gross JB (2012) The complex origin of *Astyanax* cavefish. BMC Evol Biol 12:1–12

Gross JM, Wilkens H (2013) Albinism in phylogenetically and geographically distinct populations of *Astyanax* cavefish arises through the same loss-of function *Oca2* allele. Heredity 111: 122–130

Gross JB, Borowsky RL, Tabin CJ (2009) A novel role for *Mc1r* in the parallel evolution of depigmentation in independent populations of the cave fish, *Astyanax mexicanus*. PLoS Genet 5:1–14

Hausdorf B, Wilkens H, Strecker U (2011) Population genetic patterns revealed by microsatellite data challenge the mitochondrial DNA based taxonomy of *Astyanax* in Mexico (Characidae, Teleostei). Mol Phylogenet Evol 60:89–97

Hendrickson DA, Krejca JK, Rodríguez Martinez JM (2001) Mexican blindcats genus *Prietella* (Siluriformes: Ictaluridae): an overview of recent explorations. Environ Biol Fishes 62: 315–337

Hinaux H, Poulain J, Da Silva C et al (2013) De novo sequencing of *Astyanax mexicanus* surface fish and Pachón cavefish transcriptomes reveals enrichment of mutations in cavefish putative eye genes. PLoS One 8(1):e53553. doi:10.1371/journal.pone.0053553

Hubbs CL, Innes WT (1936) The first known blind fish of the family Characidae: a new genus from Mexico. Occasional Papers of the Museum of Zoology, vol 342. University of Michigan Press, Ann Arbor, pp 1–7

Juberthie C (2000) Generalities and the diversity of the karstic and pseudokarstic hypogean habitats in the world. In: Wilkens H, Culver DC, Humphreys WF (eds) Ecosystems of the world, Vol 30. Elsevier, Oxford, pp 17–39

Kane TC, Culver DC, Jones RT (1992) Genetic structure of morphologically differentiated populations of the amphipod *Gammarus minus*. Evolution 46:272–278

Kosswig C (1965) Génétique et l'évolution régressive. Rév Quest Sci 136:227–257

Martin AP, Bermingham E (2000) Regional endemism and cryptic species revealed by molecular and morphological analysis of a widespread species of Neotropical catfish. Proc R Soc Lond B 267:1135–1141

Miller RR, Smith ML (1986) Origin and geography of the fishes of Central Mexico. In: Hocutt C, Wiley EO (eds) The zoogeography of North American freshwater fishes. Wiley, New York, pp 487–507

Mitchell RW, Russell WH, Elliott WR (1977) Mexican eyeless characin fishes, genus *Astyanax*: environment, distribution, and evolution. Special publications of the Museum Texas Tech University, 12. Texas Tech Press, Lubbock, pp 1–89

Myers GS (1966) Derivation of the freshwater fish fauna of Central America. Copeia 1966: 766–772

Ornelas-García CP, Domínguez-Domínguez O, Doadrio I (2008) Evolutionary history of the fish genus *Astyanax* Baird and Girard (1854) (Actinopterygii, Characidae) in Mesoamerica reveals multiple morphological homoplasies. BMC Evol Biol 8:1–17

Ornelas-García CP, Bastir M, Doadrio I (2014) Morphometric variation between two morphotypes within the *Astyanax* Baird and Girard, 1854 (Actinopterygii: Characidae) genus, from a Mexican tropical lake. J Morph 275:721–731

Peters N, Scholl A, Wilkens H (1975) Der Micos-Fisch, Höhlenfisch in statu nascendi oder Bastard ? Ein Beitrag zur Evolution der Höhlentiere. J Zool Syst Evol Res 13:110–124

Picq S, Alda F, Krahe R et al (2014) Miocene and Pliocene colonization of the Central American Isthmus by the weakly electric fish *Brachyhypopomus occidentalis* (Hypopomidae, Gymnotiformes). J Biogeogr 41(8):1520–1532. doi:10.1111/jbi.12309

Perdices A, Bermingham E, Montilla A et al (2002) Evolutionary history of the genus *Rhamdia* (Teleostei: Pimelodidae) in Central America. Mol Phylogenet Evol 25:172–189

Protas ME, Conrad M, Gross JB et al (2007) Regressive evolution in the Mexican cave tetra, *Astyanax mexicanus*. Curr Biol 17:452–454

Reeves RG, Bermingham E (2006) Colonization, population expansion, and lineage turnover: phylogeography of Mesoamerican characiform fish. Biol J Linn Soc 88:235–255

Reis RE, Kullander SO, Ferraris CJ Jr (2003) Checklist of the freshwater fishes of South and Central America. Edipucrs, Porto Alegre

Schmitter-Soto J-J (2016) A phylogeny of *Astyanax* (Characiformes: Characidae) in Central and North America. Zootaxa 4109(2):101–130. doi:10.11646/zootaxa.4109.2.1

Stahl A, Gross JB (2015) Alterations in *Mc1r* gene expression are associated with regressive pigmentation in *Astyanax* cavefish. Dev Genes Evol 225:367–375. doi:10.1007/s00427-015-0517-0

Stepien CA, Morton B, Dabrowska A et al (2001) Genetic diversity and evolutionary relationships of the troglodytic 'living fossil' *Congeria kusceri* (Bivalvia: Dreissenidae). Mol Ecol 10: 1873–1879

Strecker U, Bernatchez L, Wilkens H (2003) Genetic divergence between cave and surface populations of *Astyanax* in Mexico (Characidae, Teleostei). Mol Ecol 12:699–710

Strecker U, Faúndez VH, Wilkens H (2004) Phylogeography of surface and cave *Astyanax* (Teleostei) from Central and North America based on *cytochrome b* sequence data. Mol Phylogenet Evol 33:469–481

Strecker U, Hausdorf B, Wilkens H (2012) Parallel speciation in *Astyanax* cave fish (Teleostei) in Northern Mexico. Mol Phylogenet Evol 62:62–70

Trontelj P, Goricki S, Polak S et al (2007) Age estimates for some subterranean taxa and lineages in the Dinaric Karst. Time Karst:183–189

Weber A, Allegruci G, Sbordoni V (2003) *Rhamdia laluchensis*, a new species of troglobitic catfish (Siluriformes: Pimelodidae) from Chiapas, Mexico. Ichthyol Explor Freshw 14: 273–280

Wilcox TP, Garcíy de León FJ, Hendrickson DA et al (2004) Convergence among cave catfishes: long-branch attraction and a Bayesian relative rates test. Mol Phylogenet Evol 31:1101–1113

Wilhelm O, Ewing M (1972) Geology and history of the Gulf of Mexico. Geol Soc Am Bull 83: 575–600

Wilkens H (1971) Genetic interpretation of regressive evolutionary processes: studies on hybrid eyes of two *Astyanax* cave populations (Characidae, Pisces). Evolution 25:530–544

Wilkens H (1979) Reduktionsgrad und phylogenetisches Alter: Ein Beitrag zur Besiedlungsgeschichte der Limnofauna Yukatans. Z Zool Syst Evol 17:262–272

Wilkens H (1982) Regressive evolution and phylogenetic age: the history of colonization of freshwaters of Yucatan by fish and Crustacea. Texas Mem Mus Bull 28:237–243

Wilkens H (1986) The tempo of regressive evolution: studies of the eye reduction in stygobiont fishes and decapod crustaceans of the Gulf Coast and Western Atlantic region. Stygologia 2: 139–143

Wilkens H (1988) Evolution and genetics of epigean and cave *Astyanax fasciatus* (Characidae, Pisces). Support for the neutral mutation theory. In: Hecht MK, Wallace B (eds) Evolutionary biology, vol 23. Plenum, New York, pp 271–367

Wilkens H (2001) Convergent adaptations to cave life in the *Rhamdia laticauda* catfish group (Pimelodidae, Teleostei). Environ Biol Fishes 62:251–261

Wilkens H (2007) Regressive evolution: ontogeny and genetics of cave fish eye rudimentation. Biol J Linn Soc 92:287–296

Wilkens H, Strecker U (2003) Convergent evolution of the cavefish *Astyanax* (Characidae, Teleostei): genetic evidence from reduced eye-size and pigmentation. Biol J Linn Soc 80: 545–554

Yokoyama S, Meany A, Wilkens H et al (1995) Initial mutational steps toward loss of opsin gene function in cavefish. Mol Biol Evol 2:527–532

Zardoya R, Doadrio I (1999) Molecular evidence on the evolutionary and biogeographical patterns of European cyprinids. J Mol Evol 49:227–237

# Complexity of Interrelationship Between *Astyanax* Cave and Surface Fish

# 5

**Abstract**

The haplotype distribution of the *Astyanax* surface fish and the origin of its cave forms is strongly influenced by Pleistocene climatic change, due to which large-scale extinction and recolonization by the surface fish took place. Because of this history and the persisting interfertility, the relationship between the *Astyanax* strongly eye- and pigment-reduced (SEP) and variably eye- and pigment-reduced (VEP) cave populations as well as the recent surface populations is very complex. Although based on nuclear genes clustering with the phylogenetically old SEP, the Pachón and Yerbaniz cave populations are exceptional and show close relationship to both the recent surface and the phylogenetically young VEP cave fish due to mitochondrial capture. Furthermore, the variability of eye size observed in the VEP, especially in the Chica cave population, was often explained by hybridization. This contrasts with the finding that no gene flow exists in the SEP Yerbaniz and VEP Micos cave fish populations, where cave and starved surface fish co-occur. This is explained by the surface fish being washed into the caves and not being able to compete with the well adapted SEP or VEP cave fish in darkness. These observations are in accordance with Gause's Law which predicts that two ecomorphs competing for the same resource cannot coexist in the same niche. In the cave habitat the cave fish are at an advantage, which will finally lead to the extinction of the surface specimens competing for food. This process is reinforced in the *Astyanax* caves every year by periodic low food supply outside the rainy seasons. Bottlenecking obviously regularly occurs, which is demonstrated by the low variability of several genetic markers compared with the surface fish. However, mitochondrial introgression in the SEP cave populations like Pachón and Yerbaniz has nonetheless occurred. It is suggested that the introgressed modern haplotypes in the SEP Pachón and Yerbaniz cave fish derive from VEP cave populations, which, like the recent VEP Micos cave fish, were well adapted to cave life and therefore able to co-exist and also to hybridize with the SEP cave populations after the cave systems merged due to karst erosion. The recent surface fish on the one



H. Wilkens, U. Strecker, *Evolution in the Dark*, DOI 10.1007/978-3-662-54512-6_5

hand and the SEP and VEP cave populations on the other no longer hybridize in nature. In accordance with the Biological Species Concept, the SEP and VEP cave populations therefore can be denominated as a species on their own. The cave fish speciation process provides one of the rare examples of parallel speciation.

## 5.1 General Remarks

As shown before, the origin of the *Astyanax* cave forms is strongly influenced by Pleistocene climatic changes. After extinction in the cave area during cooler times, new invasions of surface fish from the south of Mexico or from glacial refuges took place during warmer periods. Furthermore, subterranean karst is generally never a static system, but is submitted to continuous and ongoing erosion processes (Espinasa and Espinasa 2015). Subterranean water flow is steadily changing, new caves are formed, and already existing as well as new ones may merge. In addition, the water table is fluctuating. Therefore, *Astyanax* surface fish as well as cave fish could continuously colonize caves that had newly originated or in which formerly existing cave fish populations had become extinct (Espinasa and Espinasa 2015).

Today, surface *Astyanax* from above ground regularly get washed underground with inflowing water during the rainy seasons. Most probably, not only surface and cave fish, but also different cave populations would have come into contact due to erosion. Therefore, an important question is to what degree hybridization between cave and surface fish as well as between different cave fish populations has occurred and influenced the evolution of the diverse *Astyanax* cave fish populations.

## 5.2 Strongly Eye- and Pigment-Reduced Cave Populations

Although strongly eye- and pigment-reduced (SEP) cave populations are building a rather uniform group (Table 4.1) characterized by features such as strongly reduced eyes and pigmentation as well as mitochondrial and microsatellite clustering, a few exceptions can be observed: according to mitochondrial analysis, the SEP Pachón and Yerbaniz cave populations are closely related to both the recent surface and the variably eye- and pigment-reduced (VEP) cave fish (Figs. 4.1, 4.4, and 4.3). This discrepancy was attributed to introgressive hybridization and mitochondrial capture (Bradic et al. 2012; Dowling et al. 2002; Strecker et al. 2003, 2004, 2012). Mitochondrial capture in general is explained by the adaptive ability of a new haplotype leading to its fixation (Qiong-Ying et al. 2012).

### 5.2.1 Pachón Cave Fish

The Pachón cave fish looks like a typically phylogenetically old cave fish and was originally described as "*Anopthichtys antrobius*" by Alvarez in 1946 (Fig. 3.8). Most interestingly, in Pachón the zygosity changed for the albino locus. It originally showed heterozygosity (Sadoglu 1955), but was homozygous in 1971 and had become heterozygous again after 1986. Between 1986 and 1988, specimens with externally visible and larger eyes were observed among the blind fish in the Pachón cave. Although no surface fish could be detected, this phenomenon was initially explained by hybridization with recent surface fish (Langecker et al. 1991) (Figs. 5.1 and 5.2). The observed range of eye size exhibited high variability, ranging from small to large eyes, and was similar to an F2 crossing. However, such distribution pattern would have afforded the exclusive matings of F1 specimens, which is rather unlikely. Among a large number of cave fish, a backcrossing between an F1 hybrid and a cave fish would have been much more probable and large-eyed specimens should not appear. Astonishingly, the observed variability vanished very quickly and was no longer found after 1996 and 2000 (Dowling et al. 2002).

Whereas it seems obvious that hybridization had occurred, it is not known which type of *Astyanax* had hybridized with the Pachón. Microsatellite studies revealed high genetic separation with no gene flow between Pachón cave and surface fish, making hybridization rather unlikely (Strecker et al. 2012; Bradic et al. 2012) (Fig. 4.3). This result is surprising because in addition to the observed variability of eye size, mitochondrial DNA (mtDNA) studies revealed that the Pachón cave population was introgressed by a haplotype also found in surface populations (Dowling et al. 2002; Strecker et al. 2003). It could be shown that the introgression occurred earlier than the 1980s because this modern haplotype was also detected in laboratory Pachón cave fish offspring from Breder's laboratory caught in the cave in the early 1940s (Strecker et al. 2004). Thus, it is evident that hybridization has occurred at least twice in the past. The observed hybridization can therefore not result from a surface fish penetrating into the Pachón cave. Rather, it is proposed that specimens of an as-yet undetected VEP cave fish population get access to the Pachón cave during extraordinary hydrological situations. This would be corroborated by transcriptomic analysis, which revealed that a significant number of polymorphic sites are shared between the Pachón cave and the surface (Hinaux et al. 2013) or the unknown VEP population, which is hypothesized to have introgressed the Pachón cave fish. The observation of specimens deriving from such an as-yet undiscovered cave fish population in stream sources of the eastern slope of the Sierra de El Abra north of the Pachón cave was recently reported (Espinasa and Espinasa 2015).

The hybridization with a VEP cave fish would explain the occurrence of the mtDNA haplotype as well as the variability and increased size of eyes found in the Pachón cave population in 1986 (Figs. 5.1 and 5.2) (Langecker et al. 1991). In laboratory crossings of the VEP Molino with the SEP Pachón cave fish, size and variability of the eye surpass that of both parental forms (see Sect. 6.21.3.3)

**Fig. 5.1** Specimens sampled in the Pachón cave in 1986 (preserved in formaldehyde, Langecker et al. 1991)

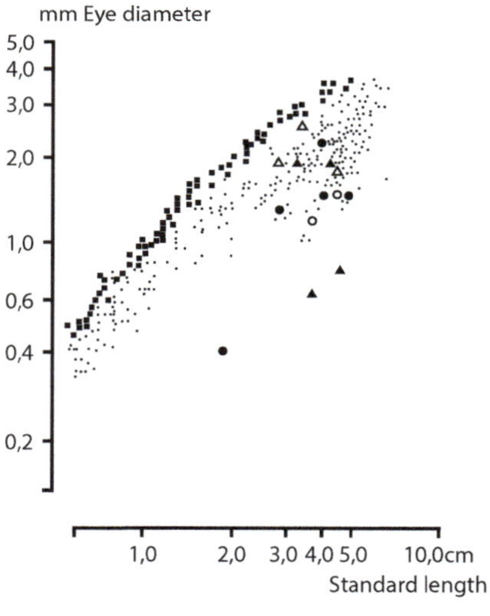


**Fig. 5.2** Distribution of eye size in the surface fish (*closed squares*), a laboratory F2 crossing between surface and Pachón cave fishes (*spots*), and the variable specimens sampled in the Pachón cave in 1986: albinotic (*white triangles*), pigmented (*black triangles*) and in 1988: albinotic (*white circles*), pigmented (*black circles*) (adapted from Langecker et al. 1991)

(Wilkens and Strecker 2003). This was explained by the recombination of eye genes having evolved in separate cave populations, which when recombined exhibit a threshold-like increase of epistatic gene effect. Such a rare combination may get lost in the next generation with the result of eye size becoming small again (Wilkens 2016, see Sect. 6.23). It is suggested that this accounts for the quick disappearance of larger eyed individuals in the Pachón cave, as observed by Dowling et al. (2002).

### 5.2.2 Yerbaniz Cave Fish

Like the Pachón cave fish, the SEP Yerbaniz cave fish was also subjected to mitochondrial capture (Strecker et al. 2004, 2012; Bradic et al. 2012). In this cave, recent surface fish are washed into the underground every year during the rainy season in enormous numbers (Elliott 2015; Mitchell et al. 1977, own observations). Strikingly, in contrast to the Yerbaniz cave fish, which appear well nourished, all surface fish sampled in the cave exhibit undernourished phenotypes, which is assumed to result from starvation (Fig. 5.3). The entrapped surface fish are starving and finally die (Mitchell et al. 1977). This observation is corroborated by microsatellite analysis which showed high genetic divergence of these two populations with only little admixture (Fig. 4.3). From this can be concluded that the observed mitochondrial capture in Yerbaniz cave fish was not by introgression of the recent surface fish but rather by a VEP cave population (Strecker et al. 2012). In contrast to the situation found in the Pachón cave, this unknown VEP population either no longer has hydrological contact with the Yerbaniz cave or completely merged with it long ago.

**Fig. 5.3** Starving *Astyanax* surface fish entrapped in the Yerbaniz cave

## 5.3 Variably Eye- and Pigment-Reduced (VEP) Cave Populations

The VEP *Astyanax* cave fish exhibit enhanced phenotypic and genotypic variability of regressive traits, which is most impressively demonstrated by eye size. Because of the variability of eye size, and particularly because of the co-occurrence with large-eyed fish, which were interpreted to be surface specimens being washed into the caves, such cave populations were called "mixed fish populations" by Mitchell et al. (1977). As surface and cave *Astyanax* can easily be crossed in the laboratory (Sadoglu 1957; Wilkens 1970a, b), it was tempting to explain this variability by hybridization (Bradic et al. 2012; Mitchell et al. 1977). However, a second hypothesis explaining the variability is often ignored: species having recently invaded caves are in general characterized by high variability of biologically functionless traits. Variability arises because such traits are no longer submitted to purifying selection eliminating regressive mutations (see Sect. 7.8) (Kosswig 1960, 1965; Wilkens 2010, 2011, 2016).

### 5.3.1 Micos Cave Fish

The eye of the Micos cave fish (Fig. 5.4) is reduced but not to the extent of SEP cave fish and the variability of its size in adult specimens found in the cave is approximately represented by the smaller half of an F2 crossing between a SEP fish and the surface fish (Wilkens and Burns 1972). The eye size of randomly bred progeny of original Micos individuals is about intermediate between the surface ancestor and the SEP cave fish (Fig. 5.5) (Wilkens 1976). Range and standard deviation of the eye size surpass those of the surface as well as of the SEP cave fish, but they are lower than in their F2 progeny. Because of its variable eye size, it has been disputed since its discovery whether the Micos cave fish hybridize with the surface fish (Wilkens and Burns 1972; Mitchell et al. 1977; Wilkens 1988; Bradic et al. 2012; Strecker et al. 2012). This was particularly focused on because during every rainy season hundreds of surface fish get washed into the Micos cave (Wilkens and Burns 1972; Mitchell et al. 1977). However, a series of studies contradicts this assumption. Micos cave and surface fish both collected in the cave can phenotypically be identified and discriminated by measuring eye size (Wilkens and Hueppop 1986) (Figs. 5.6 and 5.7). Furthermore, by comparison with surface fish from rivers it could be shown that all those collected in the Micos cave exhibit lower body condition factors. This shows that they suffer from bad nutrition. In contrast to the cave fish, they are starving whereas all Micos cave fish have higher body condition factors and do not show any sign of malnutrition (Fig. 5.8) (Espinasa et al. 2014; Wilkens and Hueppop 1986). Like in the SEP cave fish, the better body condition seems to rely on the presence of a mutated melanocortin 4 receptor allele, for which the Micos cave fish is homozygous, that contributes to elevated appetite, growth, and starvation resistance (Aspiras et al. 2015, see Sect. 6.9).

**Fig. 5.4** Large- (**a**) and small-eyed (**b**) specimens of the VEP Micos cave fish bred in daylight

The improved adaptation to cave life is also nicely confirmed by a study examining food search. Whereas the surface fish is visually orientated and therefore handicapped in absolute darkness, the cave fish rely on taste, olfaction, and the lateral line sense. Bibliowicz et al. (2013) found differences in chemosensory response and naris size between co-occurring Micos cave and surface fish (erroneously assigned as "surface-like forms") sampled in the Micos cave. According to these and our own measurements (see Sect. 6.5), the Micos cave fish possess larger naris pits and display an increased attraction to food extract odor. Because of such adaptations, surface fish cannot compete with the cave fish in the cave habitat, starve, and are finally eaten by the Micos cave fish, which was revealed by gut content analyses (Wilkens and Burns 1972).

Further proof of the Micos fish being a VEP cave fish comes from laboratory crossing analyses. Firstly, generally the manifestation of eye size exhibits epistatic gene effect (see Sect. 6.23). This means that by selectively reproducing specimens

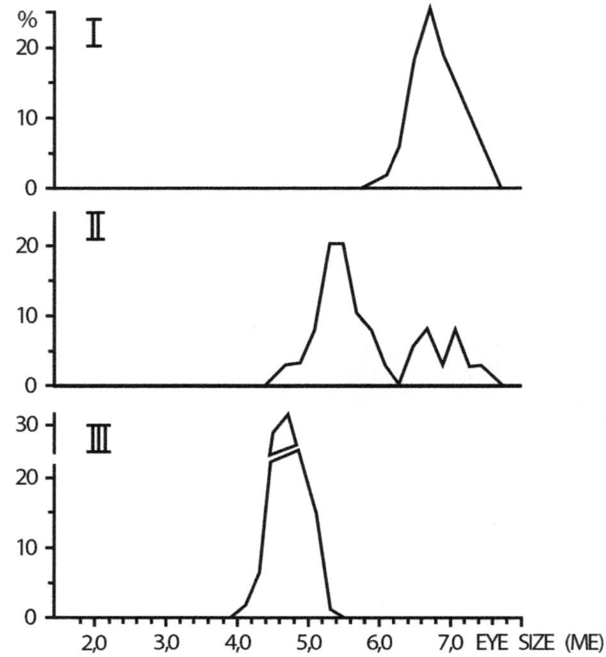


**Fig. 5.5** Distribution of eye sizes in laboratory breedings between parental Micos cave fish with large (I), intermediate (II), and small eyes (III)

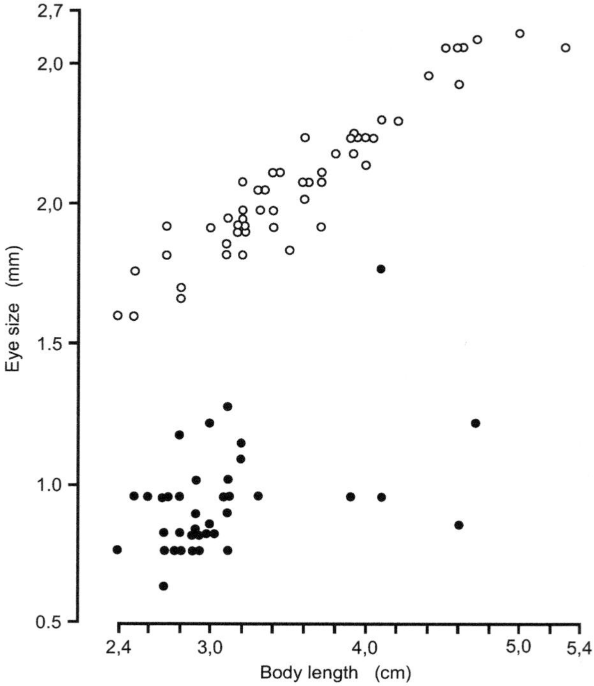


**Fig. 5.6** In the Micos cave, surface specimens flushed into it during the rainy season and true cave fish co-occur for some time and can be separated by eye size. Surface fish (*open circles*), Micos cave fish (*closed circles*) (adapted from Wilkens and Hueppop 1986)

**Fig. 5.7** Starving *Astyanax* surface and well nourished cave fish sampled in the Micos cave

with larger eyes, the distribution curve is developing bimodality because a second larger peak is formed by discontinuous increase in size (Fig. 5.5). In contrast, bimodality is not shown in F2 crossings between surface and Micos or SEP cave fish (see Sect. 6.23, Fig. 6.55a) (Wilkens 1976, 1988). In nature, such large-eyed Micos specimens forming the second peak of the bimodal distribution curve most

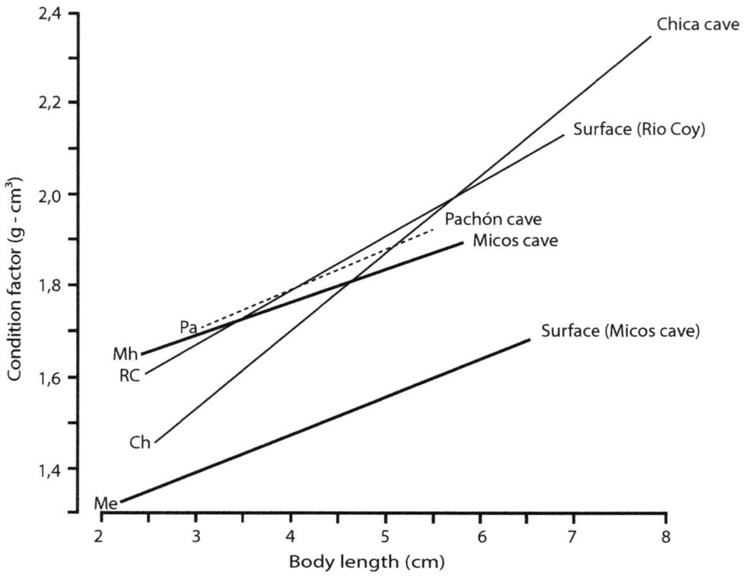


**Fig. 5.8** The correlation of the condition factors of different *Astyanax* cave and surface fish (*Ch* Chica cave fish, *Me* surface fish caught in Micos cave, *Mh* Micos cave fish, *Pa* Pachón cave fish, *RC* surface fish from Río Coy), Pachón cave fish ($n = 10, y = 1.44 + 0.09x, r = 0.27, p > 0.10$, Micos cave fish ($n = 46, y = 1.48 + 0.07x, r = 0.27, p < 0.10$), surface fish from Rio Coy ($n = 46, y = 1.31 + 0.12x, r = 0.69, p < 0.01$), surface from Micos cave ($n = 72, y = 1.15 + 0.08x, r = 0.35, p < 0.01$), Chica cave fish ($n = 24, y = 1.02 + 0.17x, r = 0.35, p < 0.01$), (adapted from Wilkens and Hueppop 1986)

probably very rarely or never occur. This is because, in the cave, mating takes place at random and the chance that two larger-eyed specimens mate is low. However, this would be required to get the combination of the necessary minimal number of eye genes for building a large eye. Secondly, it was found that the mean variability of the pupil diameter in the Micos cave fish is much lower than in the F2 crossings between surface and SEP cave fish as well as between the Micos and surface or SEP cave fish (see Sect. 6.21.3.2, Fig. 6.40) (Wilkens 1976, 1988).

Microsatellite analyses revealed high $F_{ST}$ values between Micos cave fish and the surface fish sampled in the Micos cave (Strecker et al. 2012, Bradic et al. 2012). This indicates no or only low gene flow, which was also stated by allozyme studies (Peters et al. 1975). In contrast, based on model calculations, Bradic et al. (2012) claimed that significant levels of gene flow in both directions between Micos cave and surface populations exist. This is contradictive to their own statement that the Micos cave population is able to maintain a cave-specific phenotype. It is not considered that the surface fish cannot compete with the cave fish in the cave or that the cave fish cannot compete with the surface fish in the surface creek. Other explanations for the observed "gene flow", like incomplete lineage sorting or possibly the difficult determination of cave and surface fish caught in the cave are more plausible (e.g. Bibliowicz et al. 2013) (see Sect. 6.23).

### 5.3.2 Chica Cave Fish

The first cave *Astyanax* was described as "*Anoptichthys jordani*" by Hubbs and Innes (1936) and occurs in the Chica cave (Fig. 5.9). This *Astyanax* cave is the only cave in which large bat roosts provide a rich amount of guano as food for large numbers of *Astyanax* cave fish. The amount is so enormous that even large numbers of the crayfish *Procambarus cuevaechicae* (Hobbs 1941) co-occur. Based on variable eye size, Breder (1942) divided the Chica cave fish into three groups: two of them include small eye rudiments, which are deeply sunken into the eye orbit and externally not visible ("blind") or which are visible, but in relation to their different size are "covered" or "uncovered" by tissue. The third group of eyes is characterized as "normal" and "appearing as in a river fish, irrespective of its size, which was frequently very small." However, the mean diameter of the eye of the "normal" group was significantly smaller than in surface *Astyanax*. Therefore, the classification of "normal" eyes by Breder (1942) is misleading, generating the impression of large numbers of surface fish occurring in the Chica cave. Actually, Breder (1942) shows that all the members of this group exhibit smaller eyes compared with the surface fish. Thus, these individuals are not surface fish and from the very beginning the occurrence of surface fish in the cave is left questionable. Also, Mitchell et al. (1977) did not present exact data, but just reported that surface fish occurred in the Chica cave, which may have resulted from a biased view. It was only Romero (1983), who performed measurements. He characterized the range of eye variability and found that in the pool closest to the alleged source of surface fish into the cave, it was similar to that of a backcross to a SEP cave fish. He provided no obvious evidence of the occurrence of any surface fish. The existence of surface specimens in the Chica cave may just be the result of a biased view or a rare incident. If it has occurred at all, it need not necessarily have occurred via a connection between the closest cave pool and the nearby river Tampaón, but may have originated from upstream via the creek that feeds the cave permanently with water. However, in accord with Gause's Law or the competitive exclusion principle (Gause 1934; Hardin 1960), the long-term survival of surface *Astyanax* is in doubt (see Sect. 5.4). Even if mating between surface and cave fish happen, the larvae of the hybrids would not be able to compete with the cave fish offspring. Apparently the assumption that hybridization of cave and surface fish takes place in the Chica cave relies less on the actual presence of surface fish and more on Breder's (1942) observation of a gradient of eye size increasing in the direction of the unknown alleged source of surface specimens in the deepest part of the cave. However, during the rainy season, floodings are making continuous the otherwise not completely isolated pools of the cave, which are permanently flown through by a creek (Mitchell et al. 1977). This gradient is refuted by microsatellite analysis, which revealed Chica to be a single panmictic population (Strecker et al. 2012).

**Fig. 5.9** Chica cave fish sampled in 1982 showing the variability of eye size and pigmentation

A series of findings indicate that the variability of eye size found in the Chica cave population derives from its phylogenetically young age and not from hybridization. For example, based on mtDNA studies, the Chica cave fish was found to cluster with the recent surface fish (Dowling et al. 2002; Strecker et al. 2004). Microsatellite studies revealed that the grouping of Chica is ambiguous between the phylogenetically old SEP cave fish and the recent surface fish from the northern region (Strecker et al. 2003, 2012; Bradic et al. 2012). Whereas Strecker et al. (2003, 2012) hypothesized that the VEP Chica cave fish descends from a more recent invasion, but is partly admixed by an older SEP population, Bradic et al. (2012) assumed Chica to be a SEP population hybridizing with the recent surface fish that are continuously invading the cave. They confirmed the finding of Strecker et al. (2003, 2012) that Chica could obviously not be aligned with the surface or with SEP populations. Their interpretation that Chica may even be the oldest SEP population is based on its grouping with SEP populations they found by performing a STRUCTURE analysis with $K = 2$, which will necessarily offer such a grouping. However, it is more probable that this grouping is due to gene flow that has occurred (Strecker et al. 2003) and not to descendence (see Sect. 5.4).

Furthermore, among the constructive cave-specific traits that have so far been studied in the Chica cave fish, the amount of egg yolk surpasses that of the surface fish and its mean as well as its low variability equal that of the SEP and the VEP Micos cave fish, indicating that in the Chica population this trait has already become adapted to cave life (see Fig. 6.11 in Sect. 6.11). It was also shown that the Chica cave fish, like the VEP Micos cave fish, are storing large masses of body fat in contrast to the surface fish (Hüppop 1989). This probably relies on a mutation in the melanocortin 4 receptor (*MC4R*) gene, contributing to the increase of appetite and fat storage also observed in SEP cave populations (see Sect. 6.9, Aspiras et al. 2015). These results are contradictive to Chica being a hybrid population because the variability of egg yolk content as well as of fat storage in that case would be enhanced and as high as in the F2 generations between surface and SEP cave fish (Hüppop and Wilkens 1991).

Studies of the circadian rhythm in *Astyanax* surface and cave fish showed that of the 13 amino-acid changes in Chica clock gene *PER1* compared with surface fish, 12 are identical with Pachón cave fish. Also, the light-inducible CRY1a protein reveals 5 amino-acid changes compared with the surface fish, all of which are identical between the Pachón and Chica cave fish (Beale et al. 2013). These results also support the interpretation that the Chica cave fish is not introgressed by the recent surface fish. The high similarity in these clock genes between Chica and Pachón, the two most geographically distant populations of the Sierra de El Abra, may have two explanations. The clock genes may derive from the ancestral SEP populations. They were maintained in Pachón but introgressed into the VEP Chica population. This would be in accordance with the hypothesis that in Chica introgression occurred by SEP cave fish (Strecker et al. 2003, 2012). Alternatively, the clock genes may descend from a surface ancestor preceding the recent surface populations. This ancestor seeded at least two VEP populations, Chica and a second one introgressing the Pachón population (see Sects. 5.2.1 and 5.4). By no means has the Chica cave fish with its variable eyes derived from ongoing hybridization by the recent surface fish.

It can be stated that Chica and Pachón cave fish both share genetic material of VEP and SEP cave populations but in different proportions due to introgression and not due to descendence. This also explains the unexpected grouping of the SEP Pachón with the VEP Chica populations based on next-generation phylogeography (Coghill et al. 2014). The finding that, in contrast to other SEP cave fish like Sabinos, Tinaja, and Curva wearing 11 rib-bearing thoracic vertebrae, most Chica and the Pachón cave fish possess 12 ribs furthermore corroborates this view (Dowling et al. 2002).

Genetic crossing analyses insinuate that the conspicuously increased eye size of some specimens found in Chica may also be explained by a threshold-like epistatic gene effect, which characteristically may enlarge eye size discontinuously in some specimens of the VEP cave fish populations (see Sect. 6.23, Fig. 6.55a) (Wilkens 1976, 1988, 2016). This would indicate that the few large-eyed individuals observed in the Chica cave may well be cave fish and not surface fish. It is characteristic that the necessary recombination of a minimum number of eye genes responsible for such large eyes may appear and disappear again quickly. This might explain why Breder (1942) and Mitchell et al. (1977) report large-eyed fish, whereas Romero (1983) did not find any.

### 5.3.3 Caballo Moro Cave Fish

The Caballo Moro population occurs in an unexplored cave system, the only access to which is under a large karst window. At the dimly lit bottom, specimens with small sunken eyes and a large number of fish with externally visible eyes and good vision are found (Espinasa and Borowsky 2000; Mitchell et al. 1977).

Caballo Moro cave fish exhibit high genetic divergence to all other SEP, VEP, and surface populations studied based on microsatellite analyses (Fig. 4.3) (Strecker et al. 2012; Bradic et al. 2012). They are markedly distinct in microsatellite genotype, showing almost no admixture with other nuclear genotypic clusters and are a sister group to the geographically remote populations from Southern Mexico based on Nei's DA distance. Therefore, it is assumed that they descended from another invasion wave different from all other cave populations (Strecker et al. 2012) (Fig. 4.4). The genetic separation from the recent surface populations of the northern region is also supported by mtDNA studies revealing a single unique haplotype in the Caballo Moro population (Strecker et al. 2012). It is thought that Caballo Moro, too, is a VEP cave population not containing surface fish or hybrid specimens and is probably older, descending from another invasion of surface fish from the south.

The presence of the great number of large-eyed specimens can also be explained as being caused by an epistatic "threshold-like" gene effect comparable to that found in laboratory crossing experiments of the Micos cave fish (see Sect. 6.23, Fig. 6.55a). However, the Caballo Moro population is specific, because in the lit area stabilizing selection favours the maintenance of larger eyes. Only the eyed Caballo Moro fish are still able to feed with the help of vision. Furthermore, they are able to perform the characteristic visually controlled aggressive behaviour and attack and drive away the blind fish which, due to this disadvantage, only find refuge in the dark cave environment (Espinasa and Borowsky 2000; own lab observations). It is likely that the large-eyed specimens concentrate in the lit area and also mate there. It can be shown that the two groups are moderately divergent based on microsatellite studies. Therefore, it is assumed that disruptive selection has even led to the initial stages of genetic separation of the two groups, despite the small size of the pool, representing a case of incipient sympatric speciation (Strecker et al. 2012).

### 5.3.4 Molino Cave Fish

At first glance, the Molino cave fish look like a SEP cave population, because they are pale and the eye rudiments are sunken into the orbit and are not externally visible (Fig. 5.10). However, several morphological and behavioural traits deviate from those of the SEP cave fish and exhibit a more or less intermediate position between SEP traits and those of the VEP cave fish. For example, during ontogenetic growth the eyes of the Molino cave fish surpass those of the SEP cave fish in size at all stages (see Sect. 6.21.4; Wilkens 2007). In contrast to all SEP, the Molino cave fish are still developing visual cells with intact outer segments during early ontogeny. The genetic variability of eye size is larger than in the SEP cave fish, which can be attributed to considerable heterozygosity of eye genes as is found in the other VEP cave fish populations.

The pale body colour of the Molino cave fish is caused by an albino gene (Wilkens and Strecker 2003; Protas et al. 2007). As melanin production is stopped by this mutation, crossing experiments had to be performed to show that the number of melanophores is also less reduced than in the SEP cave fish and that the brown

**Fig. 5.10** The albinotic Molino cave fish also shows slightly reduced amounts of scale guanine

gene mutation (*Mc1r*) has not occurred (Fig. 6.36) (see Sect. 6.20.2) (Wilkens 2007; Wilkens and Strecker 2003). Furthermore, in contrast to the SEP cave fish, the Molino cave population like the surface fish does not exhibit vibration attraction behaviour (VAB) (see Sect. 6.7) (Yoshizawa et al. 2010). The Molino cave fish also feeds in the same way as the surface fish and does not exhibit the lowered angle of feeding posture of the SEP cave fish (see Sect. 6.8) (Kowalko et al. 2013a).

Nuclear DNA studies revealed large genetic distances from the Molino cave fish to the surface as well as to the Micos and Caballo Moro VEP populations, which are even larger to the SEP cave fish based on microsatellites (Bradic et al. 2012) and single nucleotide polymorphisms (SNPs) (Coghill et al. 2014). The Molino cave fish possess a unique haplotype which groups with the recent surface fish from the cave area (Strecker et al. 2004). A phylogeny based on SNP showed a close relationship to geographically remote northern instead of neighbouring surface populations (Coghill et al. 2014). From the discrepancy between nuclear SNP and mtDNA data it can be concluded that either the Molino cave fish or its surface ancestor was subjected to hybridization. Thus, the Molino cave fish descends from another different and probably earlier invasion wave than the recent surface populations.

## 5.4 Role of Introgressive Hybridization

It has long been known that cave and surface *Astyanax* hybridize in the laboratory (Sadoglu 1957). Furthermore, the finding of mitochondrial capture in the SEP Pachón and Yerbaniz cave populations, where a modern haplotype is combined with nuclear markers characteristic of SEP cave populations, insinuates that hybridization must have taken place in the past. These facts and the regular occurrence of surface fish in caves like Micos and Yerbaniz have led to the idea that there is ongoing hybridization between surface and cave fish (Bradic et al. 2012; Mitchell et al. 1977; Romero 1983). However, several findings make this questionable. For example, a study of clock genes revealed that Pachón and Chica cave fish differ from the recent surface fish by a large number of mutations (Beale et al. 2013). It can furthermore be ruled out that the cave-adapted phenotype can be maintained under large gene flow from surface fish as is, for example, claimed by Bradic et al. (2012). Mitchell et al. (1977) stated that the surface fish trapped in the caves serve more as food than as mates. This was corroborated by studies of the body condition factor in the Micos cave populations which revealed that the surface fish, in contrast with the cave fish, are suffering from starvation (Wilkens and Hueppop 1986). Starving surface fish and well fed cave fish also co-occur in the Yerbaniz cave (Figs. 5.3, 5.7 and 5.8) (Mitchell et al. 1977).

The results of the study of body condition factors are in accordance with Gause's Law (Gause 1934), also called the Competitive Exclusion Principle (Hardin 1960). It predicts that two ecomorphs or species competing for the same resource cannot coexist in the same niche. In the cave habitat the cave fish are at an advantage, which will finally lead to the extinction of the competing surface form that has accidental access to this habitat during the rainy season. Even if mating between surface and cave fish occurs, which might possibly happen shortly after the rainy season when the trapped surface fish are still well nourished and fit, the hybrid offspring cannot compete with the much better adapted cave fish.

Nonetheless, hybridization has undoubtedly occurred as shown by mitochondrial introgression in the SEP Pachón and Yerbaniz cave populations which, as shown before, is very unlikely caused by the recent surface fish. Therefore, it is suggested that the introgressed modern haplotypes derive from a phylogenetically young VEP population. Such VEP populations are already adapted to cave life and are able to compete with the SEP populations. They have constructively evolved traits improving survival in the caves such as larger egg yolk content, greater larval size, and increased fat storage ability to the same extant as the SEP populations (Aspiras et al. 2015; Hüppop 1986a, b; Hüppop and Wilkens 1991). As in the Pachón population all of about 100 individuals studied contain the modern instead of a SEP haplotype (Dowling et al. 2002; Strecker et al. 2003, 2004), it is assumed that complete mitochondrial capture has occurred. This indicates that the VEP haplotype is even advantageous and favoured by selection. This introgression has also taken place in the SEP Yerbaniz cave population.

This scenario would also explain the remarkable grouping of the VEP Chica with the SEP Pachón populations found in an SNP study (Coghill et al. 2014) as well as their similarity of two clock genes (Beale et al. 2013). It is assumed that the Pachón mtDNA haplotype derives from an as-yet unknown VEP population which has the same ancestral surface population as the Chica cave fish. Because of the great geographic distance between Chica and Pachón caves, it is very unlikely that there is an underground connection allowing physical contact between them. Therefore, it is suggested that alongside Chica, this unknown VEP population has originated from the same ancestral invasion of surface fish in an isolated cave in the vicinity of the Pachón cave. This unknown population makes temporary contact under extreme hydrological conditions. Due to their already accomplished adaptations to cave life, they can coexist and hybridize with the SEP Pachón cave fish. The introgression has even occurred at least twice, as is demonstrated by the existence of the modern haplotype found in specimens already sampled in the 1940s and by the detection of a hybridization event in 1986. This would also explain the observed quick disappearance of the intermediate phenotypes with larger eyes in the Pachón cave population (Dowling et al. 2002).

For Chica, too, it is assumed that introgression has taken place, but from an unknown phylogenetically old SEP cave population (Strecker et al. 2003). Whereas Pachón is a SEP cave fish introgressed by a VEP mitochondrial haplotype, the VEP Chica is introgressed by a SEP cave fish. According to this, the close relationship between Chica and Pachón found with SNP (Coghill et al. 2015) and clock genes (Beale et al. 2013) is caused by gene exchange and not by descendence. However, the variability of eye size in Pachón and Chica has different causes: in Pachón the enhanced variability of eye size observed in and after 1986 for a transitory period of time was due to hybridization with a VEP cave population. In contrast, the characteristic continuous variability of eye size in Chica is mainly due to its phylogenetically young age and is not caused by ongoing hybridization with the recent surface fish.

## 5.5 Speciation and Taxonomy of *Astyanax* Surface and Cave Fish

### 5.5.1 Surface Fish

As was shown for cave populations like Pachón and Yerbaniz, the surface fish was also subjected to mitochondrial capture and introgression during its distributional history. Even specimens of the most distant haplotype lineages have hybridized (Hausdorf et al. 2011; Strecker et al. 2003, 2004, 2012). It was found that the distribution of nuclear genotypes is not congruent with that of the mitochondrial clades. Admixture analyses suggest there has been nuclear gene flow between populations defined by different mitochondrial clades. Gene flow also occurred between surface populations belonging to different nuclear genotypic clusters. Even the Trans-Mexican Volcanic Belt, which as a geographic barrier limits gene flow, has been crossed by different surface *Astyanax* haplotype lineages several times. In Yucatán, where obvious geographic barriers are missing, the incongruence between the distribution of nuclear and mitochondrial markers reflects random colonization events caused by inundations or marine transgressions resulting in random phylogeographic breaks (Hausdorf et al. 2011) (Fig. 5.11).

This indicates that neither the nuclear genotypic clusters nor the mitochondrial clades represent independent evolutionary units, although the mitochondrial divergences are high and in a range usually characteristic for different fish species. The incongruence between the distribution of nuclear and mitochondrial markers leading to the observed diverse genetic pattern is probably due to repeated regional extinction and recolonization of surface *Astyanax* from the South and from glacial refuges in relation to Pleistocene climatic changes. Therefore, the barcoding method generating a large number of new *Astyanax* species, as for example applied by Ornelas-García et al. (2008), is not suitable for species discrimination in *Astyanax* at all.

Since the analysis of Eigenmann (1917), the taxonomy of *Astyanax* (a genus ranging from Southern Argentina up to the Rio Grande drainage system in Northern Mexico) has not been fully resolved (Géry 1977; Reis et al. 2003; Schmitter-Soto 2016). The *Astyanax* surface fish occupies the niche of a schooling midwater predator, showing more or less similar overall morphology despite its distribution in all types of aquatic habitats like pools, rivers, lakes, lagoons, and cenotes within a large geographic area.

Problems of *Astyanax* taxonomy arise from overlapping morphometric and meristic variation, which results from the complex biogeographic, geologic, and climatic history of the distribution area as well as the occurrence in different types of aquatic habitats from large rivers to small pools. For example, based on cytochrome b analysis, the species *Bramocharax caballeroi* occurring in the isolated

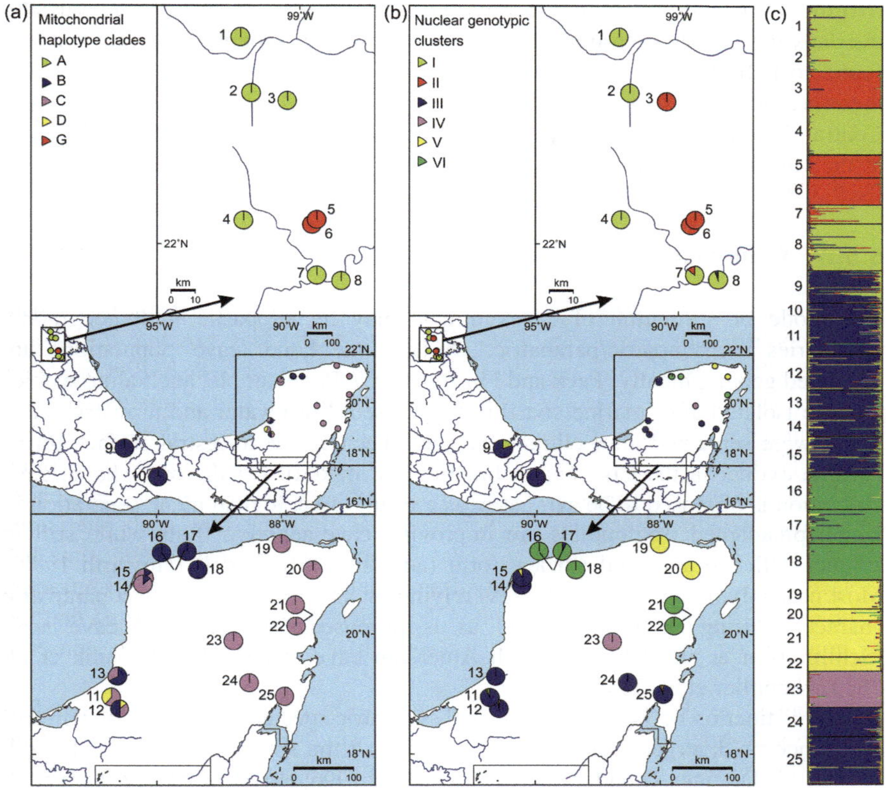


**Fig. 5.11** Geographic distribution of genetic variation in Mexican *Astyanax* populations. Numbers refer to sampling sites listed below. (**a**) Distribution of mitochondrial haplotype clades. (**b**) Distribution of nuclear genotypic clusters delimited using STRUCTURAMA. (**c**) Results of the admixture analysis with STRUCTURE for $K = 6$, *s* surface, *c* cave fish: (1) Rio Boquillas (s), (2) Arroyo Lagarto (s), (3) Pachón (c), (4) Micos creek (s), (5) Sabinos (c), (6) Tinaja (c), (7) Chica (c), (8) Rio Panuco (s), (9) Rio Papaloapan (s), (10) Rio Coatzacoalcos (s), (11) Xbacab (s), (12) Laguna Silvituc (s), (13) Rio Champotón (s), (14) Cenote Dos Marias (s), (15) Cenote Celestún (s), (16) Cenote Noc-Ac (s), (17) Cenote Dzityas (s), (18) Cenote Media Luna (s), (19) Ria Lagartos (s), (20) Laguna Leona Vicario (s), (21) Cenote Tortuga (s), (22) Cenote Actun Ha (s), (23) Laguna Chichancanab (s), (24) Valle Hermoso (s), (25) Mahajual (s) (adapted from Hausdorf et al. 2011)

Mexican Lake Catemaco as well as other species of this genus in other Central American lakes turned out to be morphotypes of *Astyanax* associated with lacustrine habitats, which is supported by the low genetic divergence between specimens of *Bramocharax* and *Astyanax* (Ornelas-García et al. 2008, 2014).

Recently in Mexico, three species of *Astyanax* surface fish have been listed (Miller et al. 2009): A. *mexicanus* (Northern Mexico to Southern Yucatán), *A. aeneus* sensu latu (Southern Mexico and Yucatán to Panama), and *A. altior*

(Northern Yucatán) (Schmitter-Soto 1998). The species A. *mexicanus* partially overlaps the distribution area of *A. aeneus* sensu latu though, and *A. altior* was found to hybridize in nature with *A. aeneus* s. l. (Schmitter-Soto 1998). It was therefore proposed to just delimit one single surface *Astyanax* species in Upper Central America (Strecker et al. 2004). This would be A. *mexicanus* (de Fillippi) as the younger or A. *fasciatus* (Cuvier 1819) as the older synonym (Fig. 4.1).

### 5.5.2 Cave Fish

The mode of speciation of cave animals may be allopatric with geographic boundaries or sympatric/parapatric. On the one hand, cave populations are separated geographically (Peck and Finston 1993). For example, anchialine species became isolated after marine coastlines regressed. Freshwater and terrestrial cave forms were separated after climate changed, because the ancestral surface forms became extinct in the cave area (climatic-relic model). On the other hand, cave speciation is proposed to be sympatric as a result of ecological niche separation in which preadapted troglophiles are improving cave-adaptive traits while still in contact with the ancestral surface form (adaptive-shift model) (Howarth 1987). Most probably, the speciation process relying on an adaptive shift has a parapatric character during an initial phase, as is proposed for a tropical cave snail (Schilthuizen et al. 2012) or North American cave salamanders (Bendik et al. 2013; Niemiller et al. 2008).

The distinction between allopatric and sympatric speciation depends on whether vicariance or divergent selection pressures drive the speciation process (Bendik et al. 2013; Desutter-Grandcolas and Grandcolas 1996; Howarth 1987; Rivera et al. 2002). *Astyanax* surface fish are diurnal midwater predators that do not actively colonise caves. This is reflected by their biogeographic distribution pattern in the Yucatán. Whereas the nocturnal troglophile catfish *Rhamdia guatemalensis* expanded through the underground cave system all over the Yucatán peninsula, surface *Astyanax* are missing in the land-locked interior cenotes and exclusively occur in a narrow coastal margin (Fig. 4.5) because there are no surface rivers they could use for distribution. In the karst area inhabited by the *Astyanax* cave fish in Northeastern Mexico, however, surface creeks and rivers exist and are captured by subterranean streams. During the rainy seasons, surface fish together with other fish species and tadpoles regularly got washed into the underground where they were trapped without any possibility of return to the surface in the same manner as today.

Despite *Astyanax* surface fish not being troglophile, in the absence of already adapted cave fish and competition it was able to invade "empty caves" and to survive in absolute darkness by the incidental combination of traits. These traits allowed it to orientate, feed, and spawn due to preadaptions such as, for example, a comparatively well developed lateral line system (see Sect. 6.5) or chemically and not visually released propagation behaviour (see Sect. 6.2). It is suggested that based both on standing variation and newly arising "de novo" mutations, selection favoured those phenotypes, which exhibited advantages for survival in the new environment. These specimens were able to outclass those which had not yet achieved such adaptations. Thus, in the beginning the speciation process may have been sympatric/parapatric because during every rainy season new specimens were washed into the underground. Some of them carried alleles improving traits important to survive and others, which did not, died in competition with the residential fish (Wilkens and Hueppop 1986). It cannot be ruled out, though, that the climatic conditions which caused the *Astyanax* fish to temporarily become extinct in the surface waters may also have played a role in the initial phase of cave colonization. During those periods, only few or no surface fish were washed into the underground. In such case it would be allopatric speciation at work, as can be observed in the well-adapted Micos and the Yerbaniz cave populations today.

Influenced by the interfertility of surface and cave fish experienced in the laboratory (Sadoglu 1955, 1957; Wilkens 1970a, b), the taxonomic status of the *Astyanax* cave populations is controversial. Originally, some of them were described as a species of their own (Alvarez 1946, 1947; Hubbs and Innes 1936). At present, it is argued in the manner of a vicious circle that because surface and cave fish may be crossed, they were belonging to one species and because they were belonging to one species, they can be successfully crossed. As the prime example of the necessary proof of the validity of the Biological Species Concept (BSC) for surface and cave *Astyanax*, the Chica cave fish population is usually quoted to demonstrate that hybridization also occurs in nature (Bradic et al. 2012; Mitchell et al. 1977).

However, as shown before, at present hybridization seems to be a very rare event between *Astyanax* surface fish on the one hand and SEP as well as VEP cave fish on the other hand in the natural caves, because surface and cave forms are adapted to different habitats (see Sects. 5.4 and 5.5) and according to Gause's law, one of them is inferior to the better adapted one in the same habitat. Therefore, based on the BSC it is suggested that surface and cave fish can be attributed the status of being two independent species.

The cave populations of *Astyanax* might represent one of the few examples of parallel speciation, which is defined as the repeated independent evolution of the same reproductive isolating mechanism (Nosil et al. 2002; Rundle et al. 2000; Schluter and Nagel 1995). The *Astyanax* cave populations provide all three criteria necessary for its denomination. First, molecular data indicate that the *Astyanax* cave populations originated during different invasions of *Astyanax* surface fish into northern Mexico (Dowling et al. 2002; Hausdorf et al. 2011; Strecker et al. 2003, 2004, 2012). At least four independent invasions of surface fish into caves are suggested by different mitochondrial clades and/or nuclear genotypic clusters. The phylogenetic independence is revealed by the parallel evolution of the regression of eyes, a process which relies on mutations in different genes in the different caves (Wilkens 1971, 2010; Wilkens and Strecker, 2003). Second, the restriction of gene flow is demonstrated for several caves where cave and surface populations are in contact. It can most impressively be observed in Yerbaniz where cave and surface populations syntopically co-occur without admixture. Third, the cave populations have separate origin, but are not reproductively isolated from one another. Lack of reproductive isolation between independently derived cave populations under natural conditions is shown by the admixture between the Chica cave population, which derives from a later more recent invasion of *Astyanax* into Northern Mexico, and a SEP cave population, which originated during the first invasion (Strecker et al. 2003, 2012). Consequently, the cave-adapted *Astyanax* populations from Northeastern Mexico should be considered as one distinct species, independently from being phylogenetically young or old.

Thus, the nomenclaturally correct name for all cave populations would be *Astyanax jordani* (Hubbs and Innes 1936), because this was the first *Astyanax* cave fish population detected and described. The original genus "*Anoptichthys*" has to be abolished, because it was incorrectly delimited owing to the fact that eyeless and pale troglobionts were formerly usually denominated a genus of their own. The scientific names for the Pachón cave fish (*Anoptichthys antrobius* Alvarez 1946) and the Sabinos cave fish (*Anoptichthys hubbsi* Alvarez 1947) are also no longer valid.

## References

Alvarez J (1946) Revisión del genero *Anoptichthys* con descripción de una especie nueva (Pisces, Characidae). An Esc Nac Cienn Biol Mex 4:263–282

Alvarez J (1947) Descripción de *Anoptichthys hubbsi* caracinido ciego de la Cueva de Los Sabinos, S. L. P. Rev Soc Mex Hist Nat 8:215–219

Aspiras AC, Rohner N, Martineau B et al (2015) Melanocortin 4 receptor mutations contribute to the adaptation of cavefish to nutrient-poor conditions. Proc Natl Acad Sci U S A 112:9668–9673

Beale A, Guibal C, Tamai K et al (2013) Circadian rhythms in Mexican blind cavefish *Astyanax mexicanus* in the lab and in the field. Nat Commun 4:1–10. doi:10.1038/ncomms3769

Bendik NF, Meik JM, Gluesenkamp AG et al (2013) Biogeography, phylogeny, and morphological evolution of Central Texas cave and spring salamanders. BMC Evol Biol 13:1–18

Bibliowicz J, Alié A, Père S et al (2013) Differences in chemosensory response between eyed and eyeless *Astyanax mexicanus* of the Rio Subterráneo cave. EvoDevo 4:25–31

Bradic M, Beerli P, García-de León F et al (2012) Gene flow and population structure in the Mexican blind cavefish complex (*Astyanax mexicanus*). BMC Evol Biol 12:1–16

Breder CM (1942) Descriptive ecology of La Cueva Chica, with special reference to the blind fish *Anoptichthys*. Zoologica 27:7–15

Coghill LM, Hulsey CD, Chaves-Campos J et al (2014) Next generation phylogeography of cave and surface *Astyanax mexicanus*. Mol Phylogenet Evol 79:368–374

Desutter-Grandcolas L, Grandcolas P (1996) The evolution toward troglobitic life: a phylogenetic reappraisal of climatic relict and local habitat shift hypotheses. Mémoires de Biospéléologie 23:57–63

Dowling TE, Martasian DP, Jeffery WR (2002) Evidence for multiple genetic forms with similar eyeless phenotypes in the blind cavefish, *Astyanax mexicanus*. Mol Biol Evol 19:446–455

Eigenmann CH (1917) The American Characidae (Part I). Mem Mus Comp Zool 43:1–102

Elliott RE (2015) Cave biodiversity and ecology of the Sierra de El Abra region. In: Keene AC, Yoshizawa M, McGaugh SE (eds) Biology and evolution of the Mexican cavefish. Elsevier, Amsterdam, pp 59–75

Espinasa L, Bibliowicz J, Jeffery WR et al (2014) Enhanced prey capture skills in *Astyanax* cavefish larvae are independent from eye loss. EvoDevo 5:35–42

Espinasa L, Borowsky RL (2000) Eyed cave fish in a karst window. J Cave Karst Stud 62:180–183

Espinasa L, Espinasa M (2015) Hydrogeology of caves in the Sierra de El Abra region. In: Keene AC, Yoshizawa M, McGaugh SE (eds) Biology and evolution of the Mexican cavefish. Elsevier, Amsterdam, pp 41–57
Gause GF (1934) The struggle for existence. Williams and Wilkins, Baltimore
Géry J (1977) Characoids of the world. TFH Publication, Neptune City, NJ
Hardin G (1960) The competitive exclusion principle. Science 131:1292–1297
Hausdorf B, Wilkens H, Strecker U (2011) Population genetic patterns revealed by microsatellite data challenge the mitochondrial DNA based taxonomy of *Astyanax* in Mexico (Characidae, Teleostei). Mol Phylogenet Evol 60:89–97
Hinaux H, Poulain J, Da Silva C et al (2013) De novo sequencing of *Astyanax mexicanus* surface fish and Pachón cavefish transcriptomes reveals enrichment of mutations in cavefish putative eye genes. PLoS One 8(1):e53553. doi:10.1371/journal.pone.0053553
Hobbs HH Jr (1941) A new crayfish from San Luis Potosi, Mexico (Decapoda, Astacidae). Zoologica 26:1–4
Howarth FG (1987) The evolution of non-relictual tropical troglobites. Int J Speleol 16:1–16
Hubbs CL, Innes WT (1936) The first known blind fish of the family Characidae: a new genus from Mexico. Occ Pap Mus Zool Univ Mich 342:1–7
Hüppop K (1986a) Oxygen consumption of *Astyanax fasciatus* (Characidae, Pisces): a comparison of epigean and hypogean populations. Environ Biol Fishes 17:299–308
Hüppop K (1986b) Comparative early life history in surface and cave fish *(Astyanax fasciatus).* Bull Zool 35(Suppl):104
Hüppop K (1989) Genetic analysis of oxygen consumption rate in cave and surface fish of *Astyanax fasciatus* (Characidae, Pisces). Further support for the neutral mutation theory Mémoires de Biospéléologie 16:163–168
Hüppop K, Wilkens H (1991) Bigger eggs in subterranean *Astyanax fasciatus.* Z Zool Syst Evol 29:280–288
Kosswig C (1960) Zur Phylogenese sogenannter Anpassungsmerkmale bei Höhlentieren. Int Rev Gesamten Hydrobiol 45:493–512
Kosswig C (1965) Génétique et l'évolution régressive. Rév Quest Sci 136:227–257
Kowalko JE, Rohner N, Linden TA et al (2013a) Convergence in feeding posture occurs through different loci in independently evolved cave populations of *Astyanax mexicanus.* Proc Natl Acad Sci U S A 110:16933–16938
Langecker TG, Wilkens H, Junge P (1991) Introgressive hybridization in the Pachon cave population of *Astyanax fasciatus* (Teleostei: Characidae). Ichthyol Explor Freshw 2:209–212
Miller RR, Minckley WL, Norris SM (2009) Peces dulceacuícolas de México. Conabio/Ecosur/Simac/Dfc, Mexico City
Mitchell RW, Russell WH, Elliott WR (1977) Mexican eyeless characin fishes, genus *Astyanax*: environment, distribution, and evolution. Special publications of the Museum Texas Tech University 12:1–89
Niemiller ML, Fitzpatrick BM, Miller BT (2008) Recent divergence-with-gene-flow in Tennessee cave salamanders (Plethodontidae: *Gyrinophilus*) inferred from gene genealogies. Mol Ecol 17:2258–2275
Nosil P, Crespi BJ, Sandoval CP (2002) Host-plant adaptation drives the parallel evolution of reproductive isolation. Nature 417:440–443
Ornelas-García CP, Domínguez-Domínguez O, Doadrio I (2008) Evolutionary history of the fish genus *Astyanax* Baird and Girard (1854) (Actinopterygii, Characidae) in Mesoamerica reveals multiple morphological homoplasies. BMC Evol Biol 8:1–17
Ornelas-Garía CP, Bastir M, Doadrio I (2014) Morphometric variation between two morphotypes within the *Astyanax* Baird and Girard, 1854 (Actinopterygii: Characidae) genus, from a Mexican tropical lake. J Morphol 275:721–731. doi:10.1002/jmor.20252
Peck SB, Finston TL (1993) Galapagos Islands troglobites: the questions of tropical troglobites, parapatric distributions with eyed-sister-species, and their origin by parapatric speciation. Mémoires de Biospéléologie 20:19–37

Peters N, Scholl A, Wilkens H (1975) Der Micos-Fisch, Höhlenfisch in statu nascendi oder Bastard ? Ein Beitrag zur Evolution der Höhlentiere. J Zoolog Syst Evol Res 13:110–124

Protas ME, Conrad M, Gross JB et al (2007) Regressive evolution in the Mexican cave tetra, *Astyanax mexicanus*. Curr Biol 17:452–454

Qiong-Ying T, SI-Qing L, Dan Y et al (2012) Mitochondrial capture and incomplete lineage sorting in the diversification of balitorine loaches (Cypriniformes, Balitoridae) revealed by mitochondrial and nuclear genes. Zool Scr 41:233–247

Reis RE, Kullander SO, Ferraris CJ Jr (2003) Check list of the freshwater fishes of South and Central America. EDIPUCRS, Porto Alegre

Rivera MAJ, Howarth FG, Taiti S et al (2002) Evolution in Hawaiian cave adapted isopods (Oniscoidea: Philosciidae): vicariant speciation or adaptive shifts? Mol Phylogenet Evol 25:1–9

Romero A (1983) Introgressive hybridization in the *Astyanax fasciatus* (Pisces: Characidae) population at La Cueva Chica. Nat Speleol Soc Bull 45:81–85

Rundle HD, Nagel L, Boughman JW et al (2000) Natural selection and parallel speciation in sympatric sticklebacks. Science 287:306–308

Sadoglu P (1955) A mendelian gene for albinism in natural cave fish. Experientia 13:394–395

Sadoglu P (1957) Mendelian inheritance in the hybrids between the Mexican blind fishes and their overground ancestor. Verhandlungen Deutsche Zoologische Gesellschaft Graz 7:432–439

Schilthuizen M, Rutten EM, Haase M (2012) Small-scale genetic structuring in a tropical cave snail and admixture with its above-ground sister species. Biol J Linn Soc 105:727–740

Schluter D, Nagel LM (1995) Parallel speciation by natural selection. Am Nat 146:292–301

Schmitter-Soto JJ (1998) Diagnosis of *Astyanax altior* (Characidae), with a morphometric analysis of *Astyanax* in the Yucatan Peninsula. Ichthyol Explor Freshw 8:349–358

Schmitter-Soto JJ (2016) A phylogeny of *Astyanax* (Characiformes: Characidae) in Central and North America. Zootaxa 4109(2):101–130. doi:10.11646/zootaxa.4109.2.1

Strecker U, Bernatchez L, Wilkens H (2003) Genetic divergence between cave and surface populations of *Astyanax* in Mexico (Characidae, Teleostei). Mol Ecol 12:699–710

Strecker U, Faúndez VH, Wilkens H (2004) Phylogeography of surface and cave *Astyanax* (Teleostei) from Central and North America based on *cytochrome b* sequence data. Mol Phylogenet Evol 33:469–481

Strecker U, Hausdorf B, Wilkens H (2012) Parallel speciation in *Astyanax* cave fish (Teleostei) in Northern Mexico. Mol Phylogenet Evol 62:62–70

Wilkens H (1970a) Beiträge zur Degeneration des Auges bei Cavernicolen, Genzahl und Manifestationsart. Untersuchungen an mexikanischen Höhlenfischen. Z Zool Syst Evol 8:1–47

Wilkens H (1970b) Beiträge zur Degeneration des Melaninpigments bei cavernicolen Sippen des *Astyanax mexicanus* (Filippi) (Characidae, Pisces). Z Zool Syst Evol 8:173–199

Wilkens H (1971) Genetic interpretation of regressive evolutionary processes: studies on hybrid eyes of two Astyanax cave populations (Characidae, Pisces). Evolution 25:530–544

Wilkens H (1976) Genotypic and phenotypic variability of cave animals. Studies on a phylogenetically young cave population of *Astyanax mexicanus* (Fillippi). Ann Spéléol 31:137–148

Wilkens H (1988) Evolution and genetics of epigean and cave *Astyanax fasciatus* (Characidae, Pisces). Evol Biol 23:271–367

Wilkens H (2007) Regressive evolution: ontogeny and genetics of cavefish eye rudimentation. Biol J Linn Soc 92:287–296

Wilkens H (2010) Genes, modules and the evolution of cave fish. Heredity 105:413–422. doi:10.1038/hdy.2009.184

Wilkens H (2011) Variability and loss of functionless traits in cave animals. Reply to Jeffery. Heredity 106:707–708. doi:10.1038/hdy.2010.105

Wilkens H (2016) Genetics and hybridization in surface and cave Astyanax (Teleostei): a comparison of regressive and constructive traits. Biol J Linn Soc 118:911–928

Wilkens H, Burns RJ (1972) A new *Anoptichthys* cave population (Characidae, Pisces). Annales de Spéléologie 27:263–270

Wilkens H, Hueppop K (1986) Sympatric speciation in cave fish? Studies on a mixed population of epi- and hypogean *Astyanax* (Characidae, Pisces). Z Zool Syst Evol 24:223–230

Wilkens H, Strecker U (2003) Convergent evolution of the cavefish *Astyanax* (Characidae, Teleostei): genetic evidence from reduced eye-size and pigmentation. Biol J Linn Soc 80:545–554

Yoshizawa M, Goricki S, Soares D et al (2010) Evolution of a behavioral shift mediated by superficial neuromasts helps cavefish find food in darkness. Curr Biol 20:1631–1636

# Regressive and Constructive Traits in *Astyanax* Surface and Cave Fish

# 6

**Abstract**

The surface and cave forms of the characid *Astyanax* have become a model system of outstanding importance in evolutionary research. It is extraordinary, because this diurnal day-active fish only exhibits minimal preadaptive traits allowing life in continuous darkness. In *Astyanax*, transition into the new environment took place abruptly and without a phase of gradual adaptation, as is usual when night-active, so-called troglophile surface forms start colonizing caves. Because of the specific characteristic of the surface fish, a large discrepancy and a large number of different traits have evolved in the cave form. The comparatively quick change into a drastically different environment is probably one of the reasons why surface and cave fish are still interfertile and can be genetically analyzed and phenotypically compared. Furthermore, cave colonization has been performed, at least in part, in parallel at different geographically separate locations and at different times. As a result, variously advanced stages of phenotypic evolution and their genetic bases can be studied.

The traits concerned can be categorized into different groups. Some, like eyes and dark pigmentation, regress because of the absence of light. Others, like the lateral line and taste sense, are submitted to directional selection to compensate for the loss of vision. Another group of traits, like altered sleep and activity phases, are ambiguous because they appear as loss and regressive in light but are constructive in permanent darkness. Entirely independent from the direct consequences of missing light, species living in temperate and subtropical caves must adapt to a continuously or temporarily low food supply.

Possibly because of regular bottlenecking events or even continuously low population density, *Astyanax* cave fish seem to tend to be K strategists with the egg yolk content enhanced in comparison with the surface fish. Because of such bottlenecking



H. Wilkens, U. Strecker, *Evolution in the Dark*, DOI 10.1007/978-3-662-54512-6_6

events, sex determination changed from polygenic in *Astyanax* surface fish to monogenic in the cave fish, based on one autosome exhibiting a gonosomal characteristic. By this, odd sex ratios or even the complete absence of one sex, which tends to occur in low population density under polygenic sex determination, are avoided.

Classical crossing analyses have revealed that regressive and constructive traits do not differ in their phenotypic manifestation. They are polygenic, and the phenotypic manifestation in the various crosses is identical. The gene effect underlying the phenotypic manifestation may exhibit an exponential increase at differing amounts in the various traits and crossings. An important role in constructive evolution is attributed to a threshold-like increase of gene effect, which is expressed after a minimum number of polygenes are recombined. By this, a disproportionate enhancement of number, size, or intensity of a specific trait is achieved, which surpasses the modifying and not genetically based environmental influence on trait expression, which hinders selection to act. The absence or presence of such genetic interaction helps determine whether the variability of regressive traits like eyes or pigmentation, exhibited by phylogenetically younger variably eye- and pigment-reduced *Astyanax* cave fish populations like Micos, is due to a more recent origin or to secondary hybridization with surface fish.

The eye is the trait best studied in *Astyanax* cave fish and provides a prime example of the genetic analysis of the evolution of complex constructive and regressive traits. Crossing experiments revealed that the eye is built by two developmental subunits, the dioptric lens subunit and the sensory retinal one. Both inherit independently from each other. It is proposed that the genes responsible for eye development divide into two groups: the first expresses around the sonic hedgehog (*shh*) genes during a first step of eye regression and determines the size of the primordial eye cup through regulation of *shh* expression. A second group of eye genes is suggested to regulate structural genes responsible for both the lens and the retina subunit. The identification of snake-specific sequence changes within the otherwise highly conserved long-range limb ZRS enhancer as being responsible of the progressive reduction of limbs from basal snake species provides a possible explanation for the complex pattern, how *shh* may be regulated, and how embryonic eye size is determined. It also provides a model of how all the other constructive and regressive traits evolve in *Astyanax* cave fish.

## 6.1 General Remarks

Caves are mostly colonized by so-called troglophiles. Such species are characteristically small-eyed and exhibit a nocturnal lifestyle. They are preadapted to life in complete darkness, and the number of traits and their degree of constructive improvement developed after the complete transition into the cave habitat is comparatively low. This was demonstrated for the cave catfish *Rhamdia zongolicensis* and *R. reddelli* in relation to their surface sister species *R. laticauda* (Hepapteridae) (Wilkens 2001) (Fig. 3.13), for the cave swamp eel *Ophisternon infernale* and its surface sister counterpart *O. enigmaticum* (Synbranchidae) (Fig. 3.19), and between the surface reef-dwelling *Ogilbia cayorum* and the closely related cave cusk eel *Typhliasina pearsei* (Bythitidae) (Fig. 3.16).

Among the cave-colonizing teleost species, surface *Astyanax* represents an exception because this large-eyed fish is diurnal and therefore, in contrast to nearly all ancestors of cave fish, cannot be characterized as a troglophile (see Sect. 6.14). This is unique among cave-colonizing species. Surface *Astyanax* did not actively invade caves, but during the rainy seasons was washed into the karstic underground with the floods of captured streams and creeks. The fact that *Astyanax* surface fish are not active cave invaders is nicely demonstrated by the distribution pattern of this species in Northern Yucatan, where surface rivers are absent and the troglophile hepapterid catfish *Rhamdia guatemalensis* has penetrated to every cenote via the vast underground system, whereas surface *Astyanax* is exclusively found in locations close to the coast, which it reached via coastal lagoons and marine transgressions (Fig. 4.5). During cave colonization in Northern Mexico, surface *Astyanax* was submitted to an abrupt change from light to continuous darkness. It was able to survive here because of a "coincidental" set of preadaptations such as non–visually released reproductive behaviour or a comparatively well-developed lateral line sense. Cave colonization was not the result of a "slow" transition, as is characteristic of troglophile species. This may explain why a higher number of quite different traits than those in troglophile species were subjected to evolutionary change.

The traits concerned can be categorized into different groups. Those directly depending on the information of light are no longer submitted to stabilizing selection and regress. The best known and most obvious ones are the eyes and dark melanin pigmentation, but behaviours such as schooling, dorsal light reaction, phototaxis, or optically released aggression are also affected (see Sects. 6.21, 6.20, 6.18, 6.13, 6.14, 6.12). Others are submitted to directional selection to compensate for the loss of vision. For example, orientation in space or food finding is changed by adaptive improvement of senses like lateral line, taste, and olfaction or behavioural traits like feeding posture (see Sects. 6.5–6.8). A third group exhibits an ambiguous character. For example, reduction of the amount of sleep may be characterized as regressive, but it becomes constructive under conditions of constant darkness (see Sect. 6.10). Entirely independent from the direct consequences of the absence of light, particularly cave species living in moderate and subtropical zones must adapt to a continuously or temporarily low food supply. In *Astyanax* cave fish, this adaptation is brought about by traits such as the enhanced ability to store fat and starvation resistance (see Sect. 6.9).

After the description of the first cavernicolous characid *Anoptichthys jordani*, the Chica cave fish, by Hubbs and Innes (1936) it was soon realized that this species was closely related to the widely spread Mexican *Astyanax* surface fish. Curt Kosswig (1902–1982), a German biologist who had started studying the genetic basis of complex traits like eyes and pigmentation by crossing surface and cave populations of isopodes from the Istrian and Slovenian karst (Kosswig and Kosswig 1940), also initiated crossing analyses of surface and cave *Astyanax* to gain insight into the genetics and evolution of complex traits in vertebrates. He could not immediately pursue his idea, though, because he had to emigrate from Germany to Turkey for political reasons in 1937 (Dzwillo 1982; Franck 2012). Nonetheless, about 20 years after the discovery of this cave fish, his Turkish student Perihan Şadoğlu performed for the first time crosses between the surface fish and fish from the Chica, Pachón, and Sabinos caves in the laboratory of Charles M. Breder Jr. at the New York Aquarium in the mid 1950s (Kosswig 1949, 1967; Sadoglu 1955, 1957) (Fig. 6.1).

Interfertility of surface fish and their cave derivatives is also found in other species. Examples are the Mexican live-bearing cave toothcarp *Poecilia mexicana* (Poeciliidae) and its surface sister species (Parzefall 1979), as well as surface and cave populations of the cyprinid *Garra barreimiae* from Oman (Wilkens, unpublished). Crossings were also performed between the surface troglophile catfish *R. laticauda* (Hepapteridae) and two of its cave sister derivatives, *R. zongolicensis* and *R. reddelli*, as well as between both cave species (Wilkens 2001). In *Astyanax*, all cave populations can be successfully crossed with surface fish and each other in the laboratory (Sadoglu 1957; Wilkens 1988; Wilkens and Strecker 2003) (Fig. 6.1) (see Sect. 5.5). Post-mating hybridization barriers as claimed by Borowsky and Cohen (2013) could not be confirmed in crossing experiments.

## 6.2 Reproductive Behaviour

In *Astyanax*, the sexes cannot be discriminated by sex-specific traits like body colour and, except during spawning, no specific and divergent sexual behaviour is exhibited. The adult males of all surface and cave populations can be discriminated from females, though, by their slightly more slender phenotype and above all by short serial hooklets they carry along the anal fin rays on both sides. These can be felt as a raw sensation when seizing the fin between two fingers (Wilkens 1972).

Caves can only be successfully colonized by species with reproductive behaviour that does not rely on visual cues. In darkness and in light, spawning in *Astyanax* begins when females ready to ovulate start slowly circling within a rather limited area around vertical structures like stones or plants. Male spawning behaviour is released after a male has by chance touched the female genital region with its snout. At that moment the male immediately starts quickly swimming around, trying to make contact with the ovulating female again. In the light, stimulated surface males are favoured by sight, because they can directly swim at conspecifics to try to find out whether they are ready to spawn. Male propagation behaviour is solely released by an as-yet unknown chemical cue secreted by spawning females, which is not perceived at distance. It is hypothesized that the chemical immediately decays in the water, but may persist for some time in the female body mucus. Possibly, as in goldfish (*Carassius auratus*), a postovulatory prostaglandin pheromone emitted through the female genital opening releases male propagation behaviour (Stacey 2003). By intersecting the olfactory nerves, it was revealed that this cue is perceived by olfaction and not by taste, because propagation behaviour was not performed by such males though they were kept together with synchronously spawning females and males (Wilkens 1972).

Surface and cave *Astyanax* have a promiscuous mating system (Wilkens 1972). The spawning process of a female may extend over several hours. Finally, all males are whirling around the ovulating female trying to come into lateral body contact with her. When this is achieved, the male and female lay their bodies side by side and turn up their ventral sides, simultaneously emitting sperm and a clutch of eggs using shivering body movements. Possibly the male anal fin hooklets, by hooking at the female body, help to concentrate the male sperm and the simultaneously yielded

**Fig. 6.1** *Astyanax* surface and Pachón cave fish, their F1- and F2-crossing hybrids

eggs in the region of the female genital opening where the external insemination takes place. *Astyanax* surface and cave fish do not exhibit parental care, and the fertilized eggs are whirled around and, because they become adhesive in the water, stick to plants or rocks, or slowly sink to the ground.

Thus, female surface and cave *Astyanax* only choose a mate during this actual spawning process, when the ovulating female selects a male to nestle with, and not by non-ovulating females, as incorrectly claimed by Plath et al. (2006).

## 6.3 Sex Determination

Sex determination is the mechanism by which sexual organisms direct gonad development towards distinct but reproductively compatible outcomes (Moore and Roberts 2013). The most widely known genetic sex determination is the monogenic male heterogametic system of most mammals, whereby one gene on the Y chromosome determines male sex. In contrast, in polygenic sex determination, multiple and independently segregating genes determine the sex. Polygenic sex determination was detected and primarily described by Kosswig (1935, 1964) for platyfish such as *Platypoecilus maculatus* and the swordtail *Xiphophorus helleri* (Poeciliidae). In both species, males with different numbers of sex genes exist. Dwarf males containing a higher number of sex genes show rapid sexual differentiation and produce more male progeny than other males. In contrast, so-called late males differentiate more slowly, become larger, and have fewer male offspring because they have fewer sex genes.

For *Astyanax*, laboratory breedings revealed a different proportion of sexes in surface and cave fish populations. In strongly eye- and pigment-reduced (SEP) (Pachón, Piedras, Yerbaniz, Curva) and the variably eye- and pigment-reduced (VEP) cave fish (Micos, Molino), male and female specimens exhibit one-to-one sex ratios (Table 6.1). In contrast, in surface fish significant deviation from the one-to-one sex ratio of male and female specimens was found in breedings, independent from their origin from Northern or Southern Mexico. Either the male or the female specimens may alternatively represent the larger group.

The unexpected odd and varying proportions of sexes found in laboratory breedings of *Astyanax* surface fish are probably the result of polygenic sex determination. It is proposed that the number of sex genes differs in the various specimens. In nature this is not disadvantageous because in the vast space of the rivers and other water bodies populated by the surface fish, the spawning females mate successively with large numbers of males. Through this participation of many specimens, the finally advantageous one-to-one ratio of sex is probably sustained, because the varying numbers of sex genes contained in different specimens are mixed at random. However, when breeding just a single pair of surface fish in the laboratory, as done for these studies, odd proportions will develop. For example, recombining one male and one female specimen, both containing few sex genes, will result in an overwhelming number of females, whereas breeding a pair with many sex genes provides the contrary.

**Table 6.1** Sex ratio in different breedings of *Astyanax* surface and cave fish

| Breedings | No. of females | No. of males | $\chi^2$ | Significance |
|---|---|---|---|---|
| *Surface fish* | | | | |
| Rio Coy 1 | 40 | 81 | 13.8926 | *** |
| Rio Coy 2 | 177 | 108 | 16.7053 | *** |
| Rio Coy 3 | 11 | 75 | 47.6279 | *** |
| Rio Coy 4 | 26 | 70 | 20.1667 | *** |
| Rio Teapao | 184 | 70 | 51.1654 | *** |
| *Cave fish* | | | | |
| Pachón1 | 20 | 25 | 0.5556 | ns |
| Pachón 2 | 17 | 13 | 0.5333 | ns |
| Pachón 3 | 12 | 14 | 0.1538 | ns |
| Pachón 4 | 20 | 23 | 0.2093 | ns |
| Piedras 1 | 66 | 58 | 0.5161 | ns |
| Piedras 2 | 34 | 33 | 0.0149 | ns |
| Piedras 3 | 20 | 12 | 2.0000 | ns |
| Curva 1 | 22 | 25 | 0.1915 | ns |
| Curva 2 | 15 | 18 | 0.2727 | ns |
| Yerbaniz | 13 | 13 | 0.0000 | ns |
| Molino 1 | 65 | 64 | 0.0078 | ns |
| Molino 2 | 8 | 13 | 1.1905 | ns |
| Micos 1 | 55 | 72 | 2.2756 | ns |
| Micos 2 | 15 | 16 | 0.0323 | ns |
| Micos 3 | 44 | 72 | 6.7586 | ** |
| Micos 4 | 17 | 15 | 0.1250 | ns |

Surface fish from Northern (Rio Coy) and Southern Mexico (Rio Teapao); cave fish populations from the Sierra de El Abra (Pachón, Piedras, Curva, Yerbaniz), Sierra de Guatemala (Molino), and Sierra de la Colmena (Micos)

Significance levels for sex ratio deviating from one to one: *ns* not significant, $**p < 0.01$, $***p < 0.001$

In contrast, the cave fish populations exhibit a one-to-one sex ratio that is suggested to rely on monofactorial heterozygote sex determination. It is proposed that this change results from and is an adaptation to the temporarily or continuously small population of the cave fish, as it is also exemplified by the diminished variability of molecular markers (Bradic et al. 2012; Strecker et al. 2003, 2012) (see Sect. 4.3.3). Temporary lower population density results from bottlenecks and at such occasions the polygenetic basis of sex determination is unfavourable. Odd sex ratios or, in the extreme, even the complete absence of one sex will occur. Thus, during their evolution *Astyanax* cave fish have changed from polygenic to monogenic sex determination to prevent odd sex combinations and provide constancy for the equal proportion of both sexes favourable in the cave environment.

Whereas for zebrafish and cichlids it is assumed that sex determination likely results from a combination of additive and epistatic effects at many loci (Beukeboom and Perrin 2014; Moore and Roberts 2013), in *Astyanax* cave fish the monogenic sex determination system has probably emerged from a quantitative shift caused by a threshold-like epistatic gene effect of sex-determining genes, as it was also found to manifest in all constructive and regressive traits in the *Astyanax* cave fish (Wilkens 2016) (see Sect. 6.23). In this way, an autosomal chromosome became gonosomal.

## 6.4 Auditory Capacities

Fish, like all vertebrates, possess an inner ear, in which two saclike structures—the sacculus and the lagena—in the pars inferior together with the enclosed otoliths, primarily function in sound reception (Popper 1970). In ostariophysians like the surface and cave forms of the catfish genus *Rhamdia* or characids like *Astyanax*, to increase hearing sensitivity the inner ear is connected with the swim bladder by the Weberian apparatus. Both act as amplifiers of sound waves that would otherwise be only slightly perceivable by the inner ear alone. One could therefore expect that, at least in some cave-living fish species, constructive improvement of auditory sensitivity has taken place in order to detect the movement of small prey.

One example of this might be provided by the cave cusk eel *T. pearsei*, in which the sacculus and its otolith, the sagitta, are enormously enlarged in relation to the other parts of the inner ear. This indicates an improvement in hearing sensitivity (Schemmel 1977) (Fig. 3.16). In amblyopsid cave fish, not only the sacculus and its otolith but also the whole inner ear are enlarged in comparison with ancestral surface fish (Niemiller and Poulson 2010; Poulson 1963; Poulson and White 1969).

Further comparative studies between closely related surface and cave sister forms were performed in the live-bearing toothcarp *Poecilia mexicana,* in which alterations in hearing sensitivity were not revealed. Only otolith morphology was changed (Schulz-Mirbach et al. 2010). In the *Astyanax* surface fish and the Pachón and Sabinos cave fish it was shown that no morphological differences exist between the inner ears of these fish (Schemmel 1967). Studies revealed, however, that both surface *Astyanax* and specimens from the Chica cave are able to hear a wider range than other fish species, but there was no evidence that the auditory capacities of the cave fish are improved (Popper 1970).

## 6.5 Lateral Line System

The mechanosensory lateral line system of fish provides a highly sensitive hydrodynamic sense that is used in a wide range of behaviours. Hydrodynamic stimuli originate from both abiotic and biotic sources. They include water currents and water disturbances caused by own body movement, prey, predators, and conspecifics (Montgomery et al. 2001, 2009). The responsible sensory organs are neuromasts, each of which consists of a sensory bud composed of support cells and of nervous hair cells, the apical ends of which are enclosed by a gelatinous cupula produced by the support cells. Depending on their location on the body, two types can be discerned: so-called superficial or epidermal neuromasts are distributed all over the body surface whereas canal neuromasts are located within subdermal canals developed within bones or bony tubes filled with gelatinous fluid. In principal, the canal system consists of a trunk lateral canal alongside each body side and several head canals. The canals are connected with the outside through pores, with one canal neuromast regularly situated between two pores.

Superficial and canal neuromasts exhibit a functional dichotomy which is not absolute, however (Montgomery, personal communication). Superficial neuromasts are low-pass filters encoding direct current (DC) and low-frequency oscillating flows, such as those generated by currents in the water surrounding the animal or movements of the animal itself, so-called noise. In comparison, canal neuromasts do not respond to DC flows, but respond to the higher frequency signals generated by other animals such as prey and mediate prey detection or orientation to a small vibrating source (Kanter and Coombs 2003; Patton et al. 2010). The functional difference between canal and superficial neuromasts is further supported by the behavioural findings that superficial neuromasts mediate rheotaxis, the orientation to water flow (Montgomery et al. 2009).

The mechanosensory lateral line system has been improved in cave fish to enable orientation in the dark and is one of the basic prerequisites of many of the fish species colonizing caves. Cave specimens may sense their surroundings using self-generated water movements, an ability termed hydrodynamic imaging (Bleckmann et al. 2004; Windsor et al. 2008). In a "head on" approach to a wall, the *Astyanax* cave fish reacted to avoid collision at a distance of about 4.0 mm. Furthermore, it was found that they possess and can develop and encode order in a spatial map (Burt de Perera 2004a, b; Sutherland et al. 2009).

The two components, lateral line canal neuromasts on the one hand and free neuromasts on the other, may evolve quite differently and independently in the dark cave environment. An example for a unique highly developed specialization of the head canal system in fish is developed in the cave cusk eel *Typhliasina pearsei* (Bythitidae) from Yucatán (Figs. 3.5, 3.16c, d) (Schemmel 1977; Wilkens 1982). It consists of a series of widened chambers separated from each other by small openings, the lumen of which is occupied by a neuromast cupula. Additionally, the canals are not enclosed in bones but the superficial parts of the chambers are transformed into thin membranes. In stagnant calm underground waters the functional combination of a velocity (cupula of canal neuromast) and a pressure receiver (membrane-covered chamber) mediates even the weakest damming phenomena for orientation in space and seems to signal approaching live prey. The exceptionally huge head of *T. pearsei* compared with the rest of the body correlates with the enlargement of the canal chambers. Optimal function of this sensory system is completed by the method of locomotion of this fish species, which by gliding almost motionlessly through the water driven by its gently undulating long dorsal and ventral fin fringes it is avoiding noise caused by its own locomotion. *T. pearsei* has also developed superficial neuromasts, which are concentrated on the head, with only a few being found on the body (Fig. 3.16a–d). These neuromasts are located on dermal protrusions and carry enlarged cupulas (Fig. 3.16c). In contrast to the highly specialized lateral line, the sense organs of chemoreception are poorly developed in this species. No taste buds were found on the body or on the filamentous pelvic fins, and the olfactory epithelium is small in size (Schemmel 1977).

In contrast, the lateral line system of the cave catfish *Rhamdia reddelli* and *R. zongolicensis* (Hepapteridae), which are closely related to the troglophile surface sister species *R. laticauda*, was not subjected to any constructive adaptation process. This can be attributed to the fact that the troglophile ancestor of these cave fish was already adapted to life in darkness (Figs. 2.7 and 3.13c–e) (Weber 1995). In the North American family Amblyopsidae, the surface swampfish still possess head canals, which no longer exist in the closely related cave fish. Instead, an extensive system of free neuromasts arranged on elevated distinct ridges is developed (Niemiller and Poulson 2010; Poulson 1963; Soares and Niemiller 2013).

In *Astyanax*, the lateral line system also consists of enclosed canal and superficial free neuromasts. The surface *Astyanax* belongs to those fish species characterized by a high number of superficial neuromasts, with four to eight of them on every scale and more developed on both sides of the head. The course and position of the lateral line canals and the location of the canal neuromasts enclosed in them do not differ between surface and cave forms except for the infraorbital canal, which may be shortened in cave fish (Fig. 6.2). The lateral canal extending along the whole body trunk side between head and tail regularly penetrates the scales through a pore. Reaching the head it divides into the occipital canal, which branches off dorsally connecting with that of the other body side, and in the preopercular canal, which descends down within the preopercle bone ventrally entering into the maxillary canal. The latter continues to the tip of the lower jaw and is enclosed by the articular and dentary bones. After the occipital canal has branched off dorsally, it divides into the supra- and the infraorbital canals which run above and below the eye orbit. Within the frontal bone, the supraorbital canal extends as far as to the premaxilla in the upper jaw, while the infraorbital canal takes a course enclosed by the bony orbitalia into the prefrontal bone (Schemmel 1967). For the infraorbital canal, partial fragmentation was observed in some cave specimens (Schemmel 1967). It can be assumed, though, that these are artefacts resulting from injuries, as was also erroneously observed for the lateral line (Wilkens 1977).

Whereas the canal neuromasts of the cave fish do not differ from the surface fish, alterations were found concerning the free neuromasts. The sensitive areas of individual neuromasts are two times larger than in the surface form and the cupulas are much longer (300 μm vs 50 μm) (Teyke 1990). Their sensitivity is enhanced when compared with the surface fish (Yoshizawa et al. 2010). This divergent development can already be observed at early life stages between 1 and 4 months (Yoshizawa et al. 2010). The neuromasts located in the head region are the longest, whereas they become progressively shorter caudally (Montgomery et al. 2001).

In surface *Astyanax* the superficial neuromasts on the head are restricted to narrow areas, which in part follow the position of the head canals (Fig. 6.2). In the cave fish, the number of neuromasts developed on the head has increased by extending into areas devoid of them in the surface fish. They cover almost all the space between the infraorbital and preopercular/maxillary head canals. In this

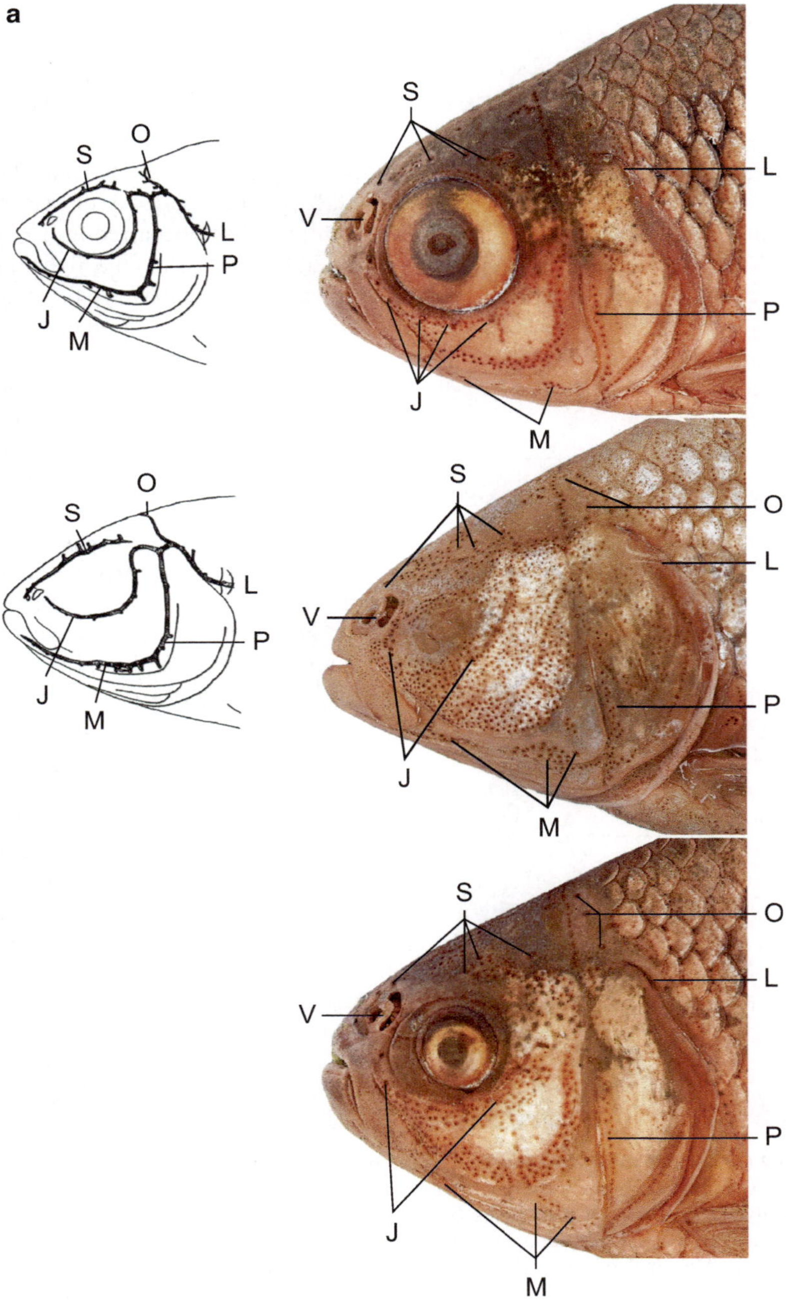


**Fig. 6.2** (**a**) Distribution of the superficial free neuromasts (*large brown dots*) and course of the head lateral line canals in *Astyanax* surface (*above*), the SEP Pachón cave fish (*middle*), and an F1 hybrid between surface and SEP Piedras cave fish (*below*).

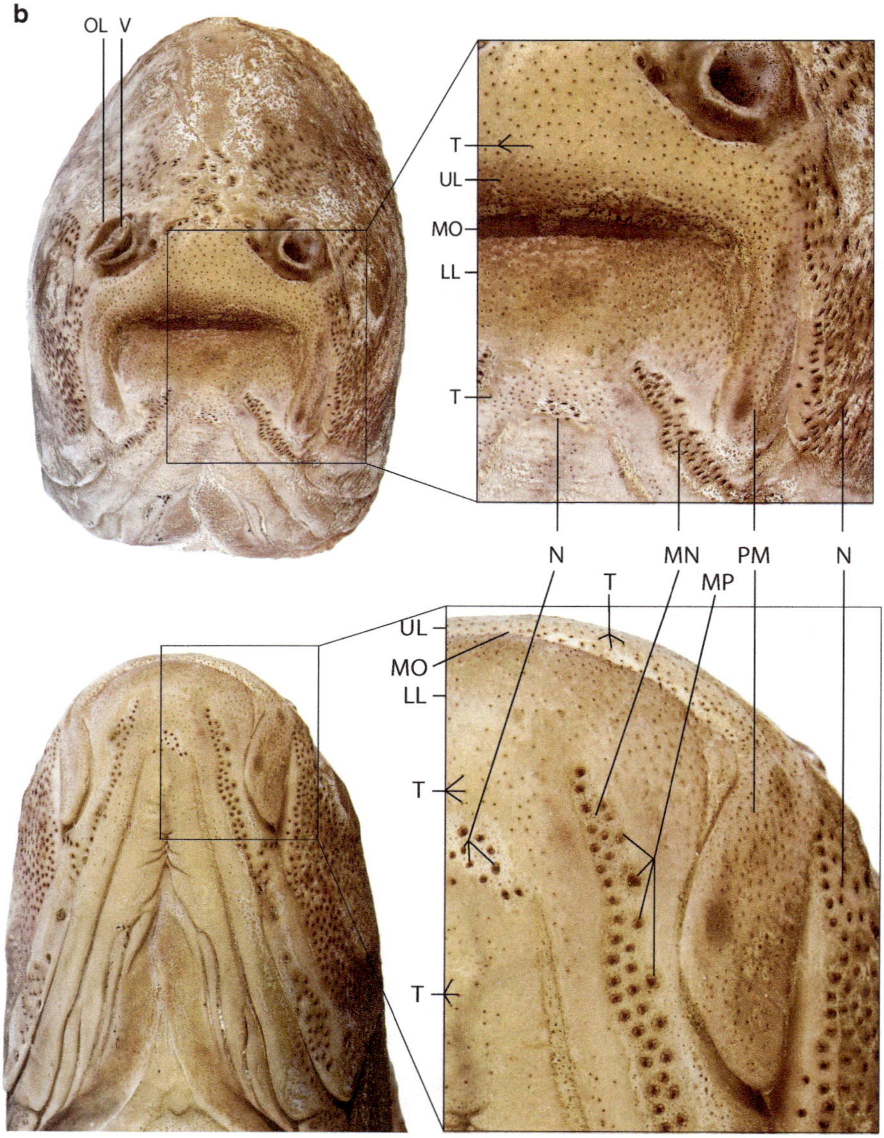


**Fig. 6.2** (continued) (**b**) Distribution of the taste organs (*tiny brown dots*) in frontal view of the head (*above*) and underneath the jaw (*below*) in the SEP Pachón cave fish. Head canals are externally recognizable by their pores (some indicated by *lines*). Staining of free neuromast cupulae and taste organ receptor areas with silver nitrate. *J* infra-/suborbital canal, *L* trunk canal, *LL* lower lip, *Mo* mouth opening, *M* mandibular canal, *MN* course of mandibular canal indicated by cupulae of free neuromasts, *MP* canal pores, *N* cupulae of neuromasts, *O* occipital canal, *Ol* nose pit with velum (*V*) over nose pit, *P* preopercular canal, *PM* premaxilla, *S* supraorbital canal, *SEP* strongly eye- and pigment-reduced, *T* taste organ, *UL* upper lip

so-called cheek area, their number has nearly doubled from about 180 in the surface fish to 350 in adult Pachón cave fish (measured at 4.5 cm standard length). Most interestingly, differences between cave fish populations were found; for example, the increase is less in the Curva (190) and the Molino (240) cave fish. In close correlation to the number of superficial neuromasts developed on the head, that following the maxillary canal below the lower jaw is also enhanced (Fig. 6.2 continued) (Wilkens, unpublished).

Although prey detection (or orientation to a small vibrating source) was found to be mediated by the canal neuromasts (Coombs et al. 2001; Kanter and Coombs 2003; for review see Montgomery et al. 2009), experimental external vibration stimuli were observed to be perceived by the free neuromasts releasing vibration attraction behaviour (VAB) (Yoshizawa et al. 2010, 2012, 2015). In contrast to most surface fish, cave fish populations were found to be able to detect the experimental vibration stimuli within a low frequency range at about 35 Hz. VAB is assumed to represent a potential foraging behaviour by which the cave fish can better locate moving prey. However, this ability is variable within and among cave populations. Whereas the SEP Sabinos and Piedras show improvement, some of the Pachón specimens lack this behaviour and it was not at all observed in the SEP Tinaja and the VEP Molino cave fish (Yoshizawa et al. 2010, 2012, 2015). The kind of prey detected by VAB in the cave is as yet unknown. Planctonic crustaceans, which should emit such vibrations, do not occur here (Elliott 2015), because green algae do not exist.

The number of superficial neuromasts developed in the cheek area in the F2 and the backcross generations between surface and Pachón cave fish range between the respective parental forms (see Sect. 6.23, Fig. 6.55e). Furthermore, it was found that the F1 hybrids between surface and SEP cave fish also exhibit an intermediate number of free neuromasts as well as intermediate abilities for the detection of water disturbances (VAB) (Fig. 6.2) (Yoshizawa et al. 2010). These results indicate a multiple polygenic mode of inheritance for the number of superficial neuromasts and also for this specific behaviour. In contrast, however, the size of the superficial neuromasts of F1 hybrids was not intermediate but equivalent to that of the Pachón cave fish (Yoshizawa et al. 2010), which can be explained by genetic epistasis. This suggests that size and number of superficial neuromasts inherit separately.

## 6.6 Taste

In fish, taste is perceived by taste buds and solitary chemosensory cells (SCCs). Taste buds are compound sensory organs that respond to food-related and tactile stimuli, whereas SCCs are single cells reacting to taste only (Hara 1994; Kotrschal 2000; Ogawa et al. 1997). In contrast to taste buds, which can only be found in the head region, they are not restricted to certain areas, but may occur all over the body surface. Their density is higher at the forehead and along the dorsal trunk (Kotrschal 1992, 1996). Taste information functionally overlaps with olfaction and both converge in the telencephalic and diencephalic nuclei of the brain.

Usually the number of taste buds is enhanced in cave fish. In species like the swamp eel *Ophisternon infernale* (Synbranchidae) (Fig. 3.19) (Parzefall and Wilkens 1972), the live-bearing toothcarp *Poecilia mexicana* (Poeciliidae) (Parzefall 1970), or the Northern blind fish *Amblyopsis spelea* (Amblyopsidae) (Poulson 1963; Poulson and White 1969), the regions covered by them are extended in comparison to their surface sister forms. In cave catfish, more taste buds probably come about by the prolongation of the barbels (Fig. 3.13) (Wilkens 2001).

In *Astyanax*, the study of the number of taste buds is incomplete and needs further investigation, because Schemmel (Schemmel 1967, 1974a, b) did not discriminate between taste buds and SCCs (Yamamoto et al. 2009; own observation). All taste organs, including taste buds and SCCs, are developed in the internal mouth cavity as well as on the upper and lower jaws with the highest densities on the labial margin surrounding the mouth opening (Fig. 6.2). Compared with the surface fish, the areas covered by taste buds have been expanded and exhibit a higher density in the cave fish. Their number has particularly increased on the upper jaw maxillary and premaxillary bone areas as well as on the ventral side of the lower jaw, where their distribution is additionally extended further caudally (Fig. 6.2). According to Schemmel's study, the SEP Pachón contains more taste organs than the SEP Sabinos, whereas the VEP Micos cave fish exhibits an intermediate position (Fig. 6.55f) (Schemmel 1967, 1974a, b).

During ontogeny of *Astyanax* larvae the taste buds start developing at 5 days post-fertilization and surface and cave forms start off with similar numbers. However, differences become notable at an age of 12 days and are significant at 22 days. At this stage the increase was 3.3- and 2.0-fold in the upper and lower jaws, respectively, in the cave fish in comparison with the surface fish. This improvement of the gustatory system is largely the result of increased taste bud density (Varatharasan et al. 2009). It is not due to the enlargement of the upper and lower jaws of the cave fish, as suggested by Yamamoto et al. (2009).

The sensory receptors make up the bulk of the cells building the taste buds (Boudriot and Reutter 2001) (Fig. 6.3). The nerve fibre plexuses of part of the taste buds contain significantly more axon profiles than in the surface fish, which was interpreted as possibly being associated with the improvement of taste. The cave fish appear to have an accelerated rate of taste bud development as well as an increased number of receptor cells per taste bud (Varatharasan et al. 2009). Unfortunately, the existence of structural histological differences between cave and surface *Astyanax* has not been sufficiently studied.

SCCs have as yet not been histologically analyzed and identified in *Astyanax* cave fish. However, as mentioned before, they may have been erroneously counted by Schemmel (1967, 1974a, b) together with the taste buds, because the number of taste buds on the chin he revealed was much higher than that actually existing (Yamamoto et al. 2009; Wilkens, own observation). The SCCs are spread all over the body surface and function by perceiving odours. They may release the fright reaction (see Sect. 6.17) when the fish are zigzagging to sample the predator's odour plume with the SCCs as described for an epibenthic fish species (Kotrschal 1996).

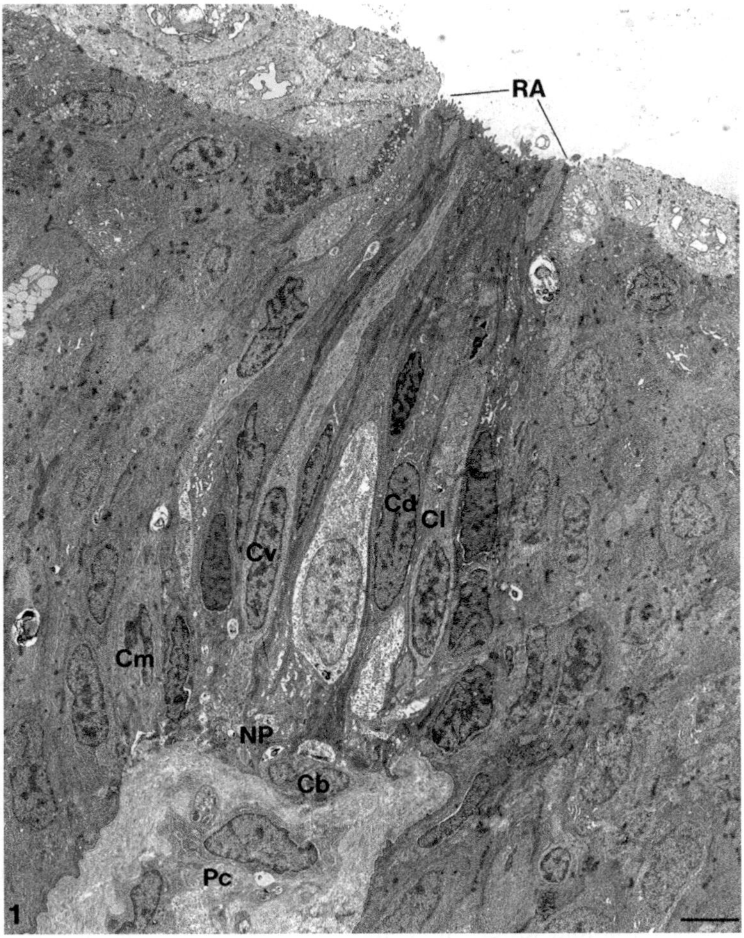


**Fig. 6.3** Longitudinal section of a taste bud of an *Astyanax* cave fish. The taste bud is situated on top of a dermal papilla (*Pc*). The sensory epithelium of the taste bud consists of elongated (*Cl*), dark (*Cd*), and dense-cored-vesicles cells (*Cv*). Their apical endings build the taste bud receptor area (*RA*). At the base there are basal cells (*Cb*) and axons of the nerve fibre plexus (*NP*). Marginal cells lie between the sensory epithelium and the unspecialized epithelium (*Cm*) (Boudriot and Reutter 2001)

The differing numbers of taste organs (buds and SCCs) between surface and cave fish are based on a polygenic system (Schemmel 1974a, b). This was confirmed by the finding of at least three different quantitative trait loci (QTL) involved (Protas et al. 2007). Crossing experiments also revealed that the increase of the taste organ number in the *Astyanax* cave fish, like all constructive and regressive traits, exhibits epistatic gene effect (see Sect. 6.23, Fig. 6.55f).

It was suggested that eye reduction in *Astyanax* cave fish might pleiotropically be driven by the increase of taste bud numbers (Yamamoto et al. 2009). However, comparison of eye size and number of taste organs in the F2 crosses did not reveal correlations between the manifestation of small eyes and large numbers of taste buds. In extreme combinations, large-eyed specimens with many taste buds as well as small-eyed ones with few of them were found. Therefore, it was concluded that the formation of eyes and taste buds is submitted to different independent genetic systems (Schemmel 1974a, b), which is also corroborated by QTL analysis (Protas et al. 2007) and lens ablation experiments (Dufton et al. 2012).

## 6.7 Olfaction

Olfaction plays an important role in fish. It may be decisive in food location, discrimination between individuals of the same or different species, defence against predators, parental care, orientation, or reproductive synchronization. Olfaction in fish is mediated by the olfactory epithelium, which is located on the lamellae. These are arranged in rosettes situated on the bottom of paired olfactory pits. Each nasal pit has an anterior inflow and a posterior outflow opening, which are partially divided by a median superficial velum directing the water flow through the anterior opening into the pit (Fig. 6.2). In synbranchid swamp eels like surface and cave *Ophisternon*, however, the nose pits are covered and closed by a dermal epithelium that separates the anterior and posterior openings (Fig. 3.19).

In *Astyanax*, the nasal pits are shallower in the cave (Pachón, Piedras, Sabinos, Micos) than in the surface fish. As a result, the olfactory epithelium might be more exposed (Fig. 6.2). In general, the nasal opening is significantly larger in the cave fish, but to varying degrees (Bibliowicz et al. 2013; Schemmel 1967). For example, in Pachón and Micos its length diameter is increased by 25%, in Curva by 17%, and in Molino by 21% compared with the surface fish (Wilkens, unpublished results). Study of 1-month-old juvenile surface and SEP Pachón cave fish using the amino acids alanine and serine resulted in a strong attractive response in both. Compared with the surface fish, the Pachón cave fish were able to perceive amino acid concentrations 105-fold lower. In contrast to the cave fish, the surface fish are only attracted for about 4 minutes. This can be explained by olfactory-driven behaviour being disrupted by vision: when they do not see any food, the surface fish will not continue searching for it (Hinaux et al. 2016).

Crossings revealed that the increase of the nasal pit size in the *Astyanax* cave fish is polygenic and, like all constructive and regressive traits, exhibits an epistatic gene effect (see Sect. 6.23).

## 6.8 Feeding Behaviour and Food Uptake

Usually cave fish derive from troglophile surface species, which are night active and therefore well adapted to life in light-poor environments. In contrast, the surface *Astyanax*, as a rare exception, is active during the day. It catches prey and food particles in the three-dimensional open water space by visual orientation

(Parzefall 1983). In caves, therefore, it has changed its behaviour and in darkness is restricted to a two-dimensional feeding approach, by solely being able to pick up food particles from the ground (Hüppop 1987). Under these conditions, the surface fish take a steep-angled feeding posture of approximately 74° (Kowalko et al. 2013a, b) to 82° (Schemmel 1980) between body axis and bottom (Fig. 6.4). This feeding posture is disadvantageous, because the swimming process is halted while food items are taken up.

The *Astyanax* cave fish mainly search for food on the ground but also feed at the water surface using an additional two-dimensional area (Hüppop 1987). In contrast to the surface form, their feeding posture on the ground is characterized by a lower angle. As a result, they are able to continuously swim, search, and swallow food particles, which in the caves are mostly found on the bottom (Fig. 6.4). Schemmel (1980) observed an average angle of about 55.7° between body axis and bottom when picking up food from the ground, which was the same in the SEP Pachón, Piedras, Yerbaniz, and Sabinos cave fish. In principle, this was confirmed by the studies of Kowalko et al. (2013a, b), although with slightly differing results, with angles of 38° in Pachón and 49° in Tinaja found. Most remarkably, the VEP Molino cave fish exhibits an angle of 66°, which is in between the surface and the SEP cave fish angles (Kowalko et al. 2013a, b) (Fig. 6.5).

In the laboratory, under darkness, the cave fish reacted much faster to food offered to them than the surface fish did. A comparable observation was made when just food odour was offered to Micos cave and surface fish caught in the Micos cave. In this study, only the cave fish started food searching behaviour and it was suggested that this was due to improved olfaction resulting from the larger naris size (Bibliowicz et al. 2013). Possibly, in the predominantly visually orientated surface fish, olfaction does not play that much of a role in food finding in its lit environments. This was confirmed by studying 1–month-old juveniles (5–6 cm body length) of the surface and the SEP Pachón cave fish, which revealed that the amino acids alanine and serine resulted in a strong attractive response in both

**Fig. 6.4** *Astyanax* surface (**a**) and cave fish (Pachón) (**b**) exhibit different angles of feeding posture when picking up food from the ground

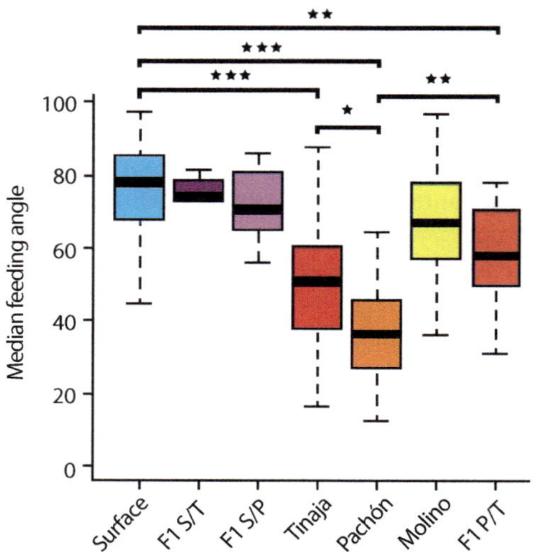


**Fig. 6.5** Feeding angle in the *Astyanax* surface, the Pachón, Tinaja, and Molino cave fish, and their crossings. *F1 S/T* surface × Tinaja, *F1 S/P* surface × Pachón, *F1 P/T* Pachón × Tinaja, $*p < 0.05$, $**p < 0.01$, $***p < 0.001$ (adapted from Kowalko et al. 2013a, b)

(Hinaux et al. 2016). However, in the light, in contrast to the cave, the surface fish are only attracted for about 4 minutes (see Sect. 6.7).

Furthermore, specific hectic food searching movements, performed by single cave fish specimens having detected food particles, immediately stimulate food searching behaviour of other cave fish nearby. In darkness, the surface fish start such behaviour much later (Hüppop 1987; Lüling 1954; Parzefall 1983).

When kept together in darkness in laboratory experiments, adult cave fish find 80% of food particles at the bottom whereas surface fish are only successful at finding 20% (Hüppop 1987). Field studies in the Pachón and the Micos cave revealed that, when small clay balls mixed with food of 5 mm in diameter are offered, the falling food balls released a higher swimming activity within a radius of 0.80 m. A rising number of cave fish from the wider periphery swimming by chance into this area began concentrating there. They started searching for food near the bottom or at the water surface. Little reaction was exhibited, however, when just clay balls of the same size were introduced, which indicates the importance of olfaction in food finding (Fig. 6.6) (Parzefall 1983). When the above-mentioned clay balls mixed with food were introduced into rivers and pools, all surface specimens (which detect the falling balls by vision) leave the school and concentrate around the food ball trying to catch particles. Thus, in surface fish active feeding behaviour is primarily visually released, but is only continued after the food particles have directly been checked by olfaction or taste. Like in the cave fish, feeding behaviour ceases when pure clay is offered.

Under the premise of food scarcity, it is often claimed that the *Astyanax* cave fish are food addicts and because of this they continuously swim in search of food

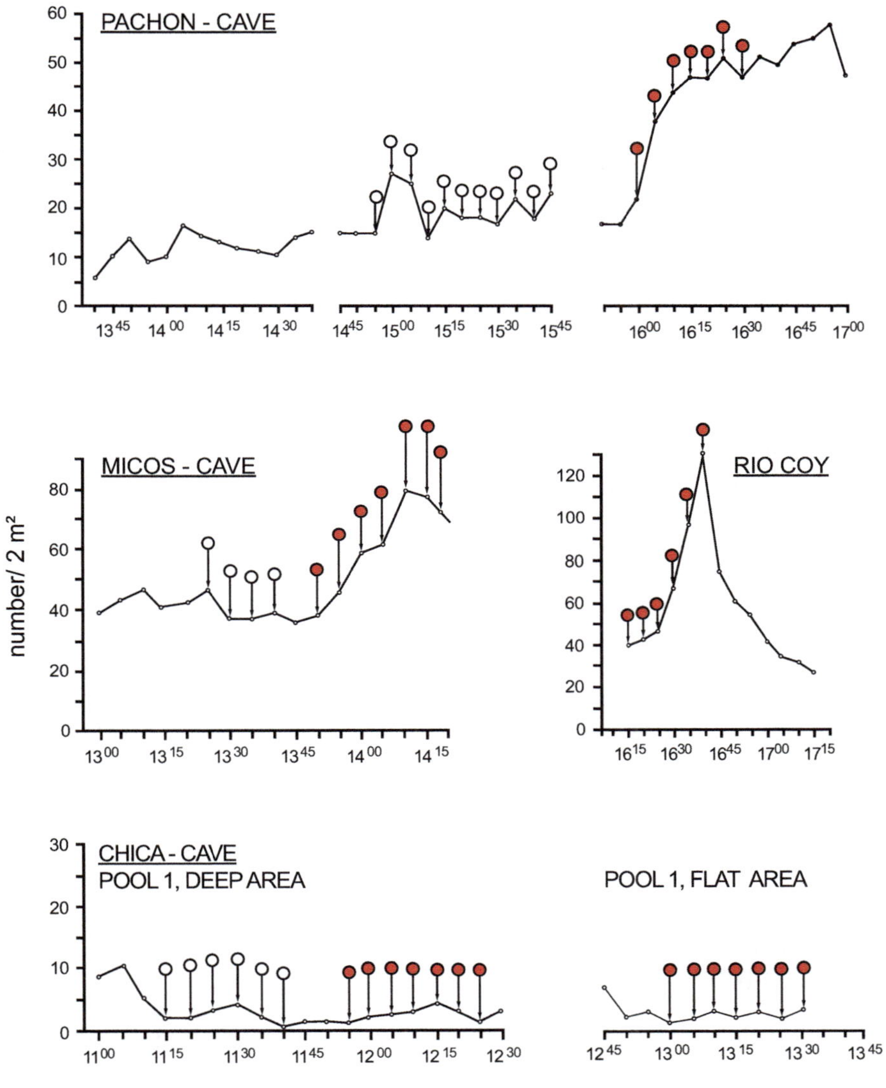


**Fig. 6.6** Density of fish in the Pachón, Micos, and Chica cave fish as well as in a surface river fish (Río Coy) before, during, and after introduction of balls of clay mixed with food (*red*) or without food (*empty circles*) within a 2 $m^2$ area (adapted from Parzefall 1983)

(Elipot et al. 2013). However, comparing surface and cave fish shows that they both show two different behavioural phases. In the first phase, low activity is characterized by slowly swimming within schools or shoals in the surface fish or within a specific home range in the cave fish. This behaviour is exhibited while the surface fish (based on vision) and the cave fish (relying on chemical senses) are both searching for food. In the second phase, high activity is released in both after a food source has been identified.

During the field experiments it was observed that neither the surface nor the Micos and the Pachón cave fish became satiated. Most cave fish even continued searching in the test area for hours. However, different observations were made in the Chica cave fish, which did not react to either pure or "mixed with food" clay balls (Parzefall 1983). This shows that the Chica fish are satiated due to the high food offer provided in the Chica cave by bat guano. This is not in accordance with the hypothetical assumptions that the cave fish were food addicts and would be insatiable (Aspiras et al. 2015; Elipot et al. 2013). Like every other species, the cave fish will probably stop feeding when saturated.

Among the above-mentioned feeding traits, only the change in the feeding posture has been genetically studied. A polygenic basis exhibiting epistatic gene effect was revealed (see Sect. 6.23, Fig. 6.55c) (Kowalko et al. 2013a, b; Schemmel 1980). Whether the genes regulating feeding posture are the same in the different cave populations is not finally resolved. The F1 cross between Sabinos and Pachón fish does deviate from the parental forms (Schemmel 1980). In contrast, the F1 cross between Pachón and Tinaja was intermediate and ranged from surface-like to cave-like feeding postures (Kowalko et al. 2013a, b). This F1 cross was significantly different from surface and Pachón fish, but not from Tinaja fish. According to these data, the genetic basis of feeding posture is suggested to have evolved independently in these two cave populations and it is suggested that several distinct QTL representing different loci are regulating feeding angles in the Pachón and Tinaja cave fish populations. Furthermore, no genetic correlations were detected between feeding posture on the one hand and morphological features like distribution and number of taste buds on the chin, altered craniofacial morphology, or eye loss on the other, either by classical crossing experiments or by QTL mapping (Kowalko et al. 2013a, b; Wilkens 1988, 2010).

## 6.9 Metabolic Adaptation to Permanent or Periodic Low Food Supply

The caves inhabited by the *Astyanax* cave fish are situated in the subtropical region, where almost all depend on periodic energy supply during the rainy season. At that time, large amounts of organic debris as well as aquatic surface species are washed into the underground. A well known exception among the *Astyanax* caves is the Chica cave, in which large masses of guano are introduced by thousands of roosting bats providing a high energy input all year long. Here, numerous specimens of the fully eyed cambarid surface crayfish *Procambarus acutus cuevachicae* even coexist with the cave fish (Mitchell et al. 1977; Wilkens and Burns 1972).

Because of the permanently or temporarily low food supply in many caves, diverse adaptations to these circumstances have been acquired by cave-dwelling animals (Hüppop 2000, 2012). The most prominent adaptations are improved starvation resistance, lowered metabolism, and enhanced fat storage. Improved

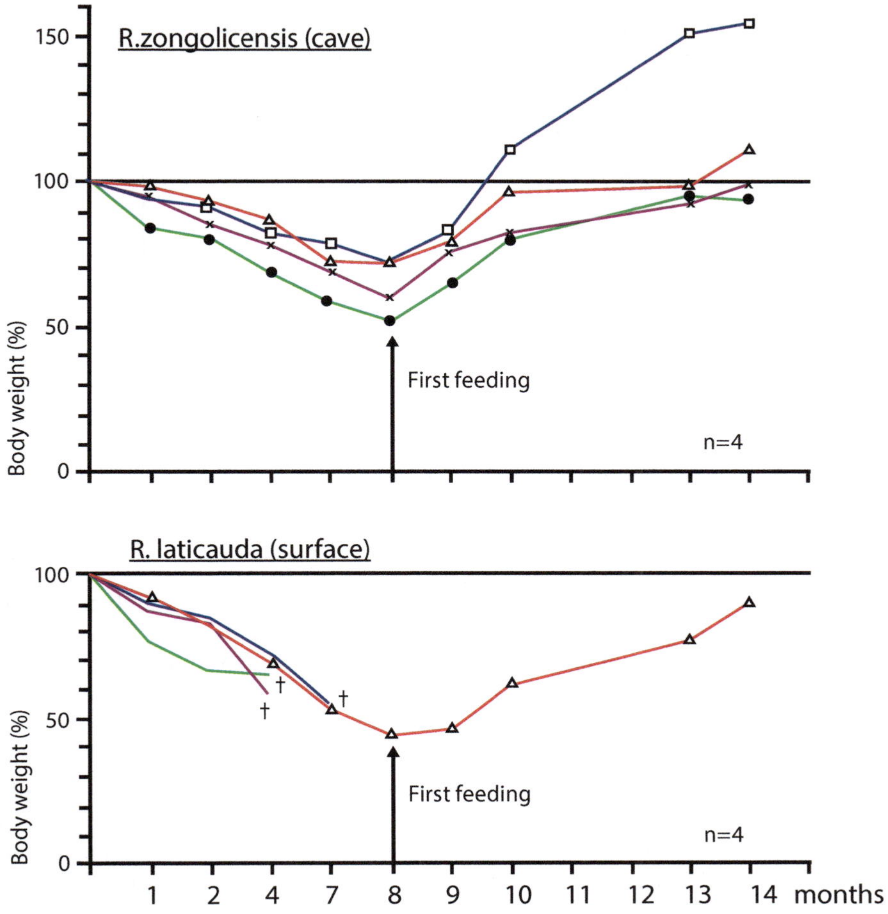


**Fig. 6.7** Development of body weight in the cave catfish *Rhamdia zongolicensis* and its surface sister species *R. laticauda* during and after a starvation phase

starvation resistance occurs in the cave catfish *Rhamdia zongolicensis*, specimens of which, due to enhanced fat deposits, are able to survive during longer starvation periods than their closely related surface sister species *R. laticauda*. The cave fish also recover more quickly (Fig. 6.7) (Wilkens 2001). The study of starvation resistance in *Astyanax* revealed that the Pachón, Tinaja, and Molino cave populations lost only half as much weight as individuals from the surface population during a 2-month fast (Fig. 6.8a). This is concordant with the finding that only after a starvation period of almost half a year does the condition factor of individual Pachón cave fish fall below that of well-fed *Astyanax* surface fish (Hüppop 2000). The study of body fat content of the Micos, Pachón, and Chica cave fish revealed that it surpasses that of the surface fish (Hüppop 1986a, b, 1989). The highest mean values were found in Pachón cave fish with 34.9% fat in wet body mass compared with

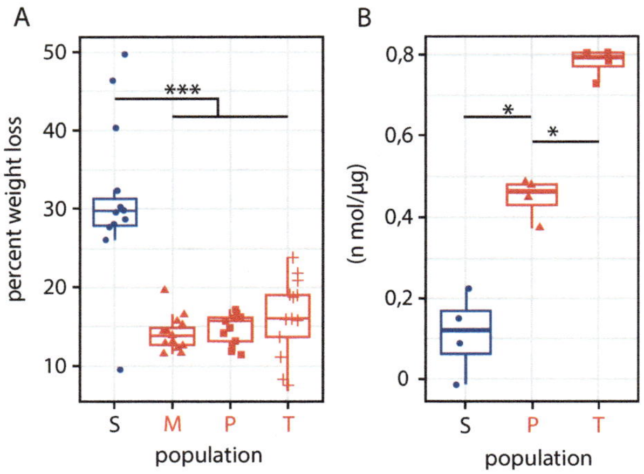


**Fig. 6.8** Adaptation of *Astyanax* surface (*S*) and cave fish (*M* Molino, *P* Pachón, *T* Tinaja) to starvation. (**a**) Percentage of weight loss after 60 days of starvation (***$p < 0.0001$, Tukey honest significant difference [HSD] test). (**b**) Total triglyceride content/protein in adult fish (*$p < 0.05$, Tukey HSD test) (adapted from Aspiras et al. 2015)

8.9% in the surface form. Consistent with this observation, the triglyceride fat content was increased in regularly fed adult Pachón and was even higher in the Tinaja cave fish compared with the surface fish (Fig. 6.8b). The only exception was observed in the Tinaja cave fish, in which fat storage takes place in the greatly enlarged liver (Apiras et al. 2015).

Food consumption of individuals from the cave and surface populations was studied to reveal how cave fish acquire increased fat content. Differences in appetite were studied by measuring the amount of freshwater oligochaetes (*Lumbriculus variegatus*) consumed over a certain period in cave and surface fish (Aspiras et al. 2015). The results suggest that there are significant differences in appetite regulation between populations. For example, although Pachón cave fish reach higher fat levels than surface fish, a difference in their appetite from their surface counterparts was not detected (Fig. 6.9a–c). Furthermore, the appetite of well fed Tinaja cave fish was not significantly different from that which surface fish reached after 3 weeks of fasting (Fig. 6.9a). Neither Pachón nor Tinaja cave fish appetites were affected by a 3-week fast, whereas surface fish that have fasted display a significant elevation in appetite (Fig. 6.9a). However, the Tinaja cave fish exhibit the highest appetite among all populations (Fig. 6.9a–c). These data suggest the possibility that there are significant differences in appetite regulation between populations.

Study of F1 hybrids between Pachón and Tinaja cave fish as well as between surface and Tinaja cave fish revealed that both exhibit appetites similar to Pachón and surface fish during a fed state. This suggests the excessive Tinaja appetite to be a recessive trait (Fig. 6.9b–c). The improved ability of the Pachón cave fish to store fat, as measured by its mean routine oxygen consumption rate over 24 hours ($Vo_2$), is intermediate in the F1 crossing and shows a slightly bimodal or at least left-

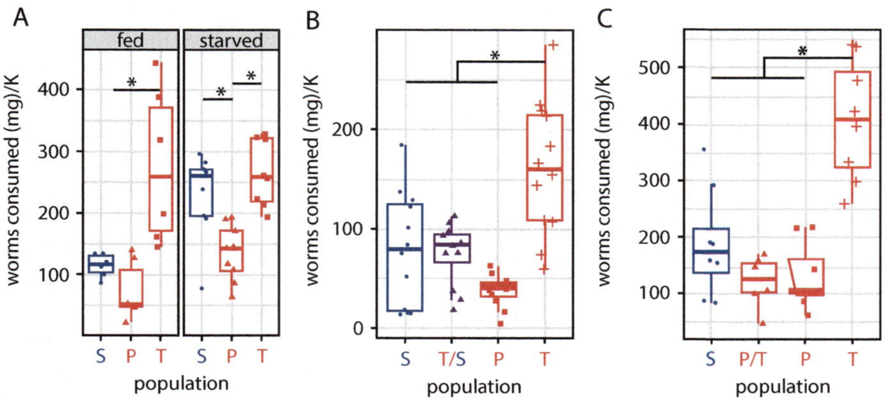


**Fig. 6.9** Differential appetite regulation. Comparison among fed and starved *Astyanax* surface (*S*), Pachón (*P*), and Tinaja (*T*) cave fish over a 3-wk period (**a**), among fed surface (*S*), Tinaja/surface F1 hybrids (*T/S*), Pachón (*P*), and Tinaja (*T*) fish over a 36-h period (**b**), and among fed surface (*S*), Pachón/Tinaja F1 hybrids (*P/T*), Pachón (*P*), and Tinaja (*T*) cave fish over a 48-h period (**c**). *K* condition factor (adapted from Aspiras et al. 2015)

oblique distribution in the F2 crossing between the surface and the Pachón cave fish, insinuating epistatic gene action (see Sect. 6.23) (Hüppop 1989).

A candidate gene approach led to the identification of coding mutations in conserved residues of the melanocortin 4 receptor (*MC4R*) gene in the Pachón, Molino, Arroyo, Yerbaniz, Tinaja, Piedras, Micos, and Sabinos cave populations, which is assumed to contribute to the increase in appetite and starvation resistance (Apiras et al. 2015). Most notably, the mutated *MC4R* is heterozygous in the Pachón cave fish and also present at very low frequency in the surface fish. This corroborates the view that standing variation may have played an important role in the initial phase of cave colonization by surface *Astyanax*.

Besides fat storage, starvation resistance, and increased appetite, lowering of metabolism is another adaptation made by cave species to survive during starvation periods. Measurements of oxygen consumption in total darkness in an absolutely silent room (camera silens) over a 24-hour period after 2 days of acclimatization revealed no difference between *Astyanax* surface fish and the Pachón cave fish, when calculating fat-free body mass (Hüppop 1989). These conditions were necessary to allow comparisons because of a different maintenance metabolism in specific body tissues, which in fat tissue is much lower than in nerve tissue, in which it is the highest (Ball and Jungas 1965; Hüppop 1989; Moran et al. 2015).

In contrast to the observation over a 24-hour period, and not considering the before-mentioned caveat (Hüppop 1989), measurements of oxygen consumption in darkness over a longer period of 7 days showed high energy savings (Moran et al. 2014). In constant darkness, the surface fish exhibited a circadian rhythm in metabolism with an increase in oxygen demand during the subjective daytime, whereas cave fish did not. It was concluded that the lack of circadian rhythm in metabolism leads to a 27% energy savings for Pachón cave fish compared with the surface fish in their natural photoperiods. Under constant darkness, the Pachón cave fish will consume 38% less energy than the surface fish. Moran et al. (2014) suggest that the lower energy demand can be explained by the fact that the metabolism in the surface fish is still subjected to circadian rhythmicity with an increase in oxygen consumption during the subjective daytime, whereas because of the lack of circadian activity in the cave fish, no increase occurred. Most notably, the metabolic cycle was uncoupled from locomotor activity.

As neural tissue exhibits the highest energetic costs of maintenance metabolism, it was often claimed that eye regression in cave fish would be driven by energy savings, particularly because the reduction of the visual system consisting of the eye and the tectum opticum are correlated. Based on experimental studies, the cost of vision was calculated to be 15% of resting metabolism for a 1 g larval fish, decreasing to 5% in an 8.5 g adult fish as relative eye and tectum size decline during growth. For the SEP Pachón and the VEP Micos cave fish, it was shown that the loss of the visual system substantially lowered the amount of energy expended on neural tissue, in particular for juvenile fish. The cost of neural tissue contained in eyes and the whole brain for a 1 g surface fish represented 23% of resting metabolism, whereas for Micos and Pachón cave fish the cost was 13 and 10%, respectively (Moran et al. 2015).

## 6.10 Overall Sleep and Activity Patterns

Circadian sleep and activity patterns were studied in juvenile fry of *Astyanax* surface fish, the SEP Pachón and Tinaja as well as the VEP Molino cave fish over a period of 48 hours (Duboué et al. 2011). Sleep profiles were based on 60 seconds of inactivity as the criterion for sleep. It was revealed that the surface fish are diurnal, sleeping at night and largely awake during the day, but with some sleep toward the middle of the day (Fig. 6.10a).

The three cave populations were found to have diverged from the surface sleep pattern and show a drastically reduced sleep phenotype both in the day and at night time (surface: 800 min vs cave: 110–250 min per 24 h) (Fig. 6.10b). The phenotype of reduced sleep has independently evolved in the cave fish as shown by the study of the SEP Pachón and Tinaja and the VEP Molino cave fish (Duboué et al. 2011). To differentiate between true sleep and reduced activity, quantification of activity per waking minute (velocity) was performed and revealed that the active waking velocities of Pachón and Tinaja but not Molino cave fish exceed those of the surface fish. However, Pachón and Tinaja are much more active both day and night compared with surface fish and Molino cave fish (Fig. 6.10c).

Sleep efficiency is generally characterized by the number and length of undisturbed bouts or periods. Whereas the number of nighttime bouts did not differ significantly between surface and cave fish, during the daytime the Pachón and Tinaja cave fish exhibited fewer bouts than the surface fish (Fig. 6.10d). Both

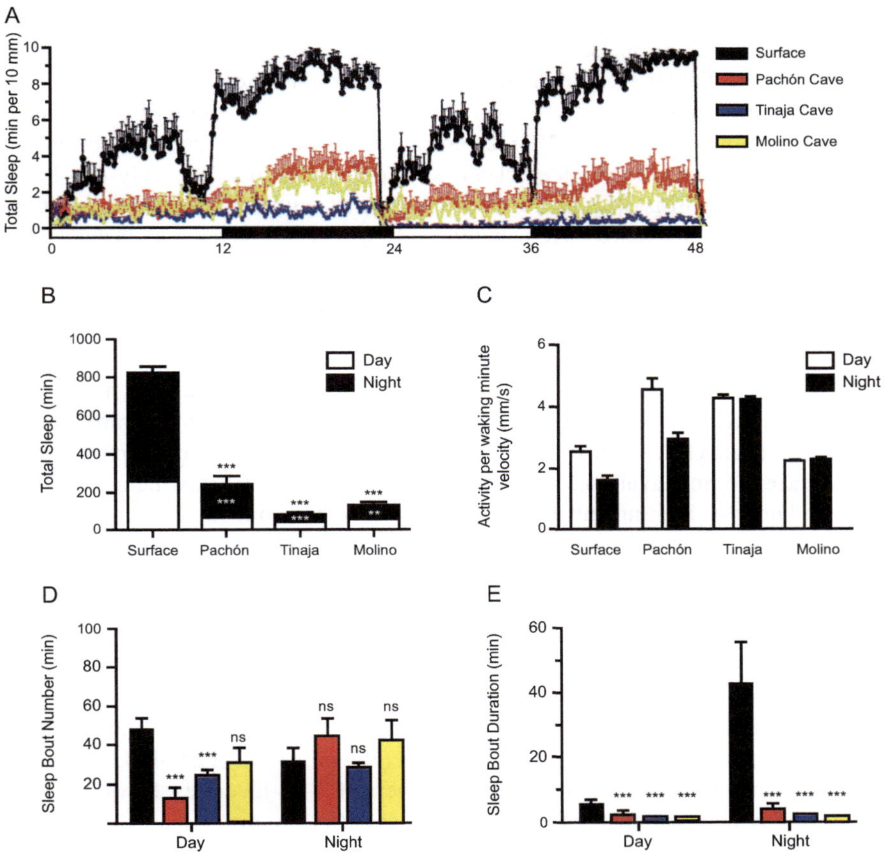


**Fig. 6.10** Amount of sleep in 21- and 24-day-old larvae of the *Astyanax* surface fish and three independently evolved cave populations. (**a**) Sleep profiles graphed as number of minutes of sleep per 10-min period over 48 h, (**b**) amount of sleep per 24 h, (**c**) average waking activity, (**d**) average number of sleep bouts, (**e**) average bout duration, *asterisks* represent significance relative to surface fish (adapted from Duboué et al. 2011)

daytime and nighttime bout lengths were shorter in all cave populations studied compared with the surface fish (Fig. 6.10e).

The studies dealing with the resting and activity patterns performed with larval fry could be stated for adult cave fish too (Yoshizawa et al. 2015). Crossing analysis revealed that the difference in total sleep between surface and cave fish has a polygenic basis (see Sect. 6.23, Fig. 6.55d) and that the genes responsible for the cave fish sleep phenotype are inherited independently from those underlying eye and pigmentation reduction (Duboué et al. 2011). It is claimed that changes in adrenergic signalling may underlie the reduced amount of sleep in *Astyanax* cave fish (Duboué et al. 2012).

Sleep is a homeostatically regulated physiological state marked behaviourally by prolonged periods of quiescence and a reduced responsiveness to external stimuli (Campbell and Tobler 1984). In mammalian species, the amount of sleep was found to vary greatly and primarily reflects ecological constraints acting on total sleep time (Capellini et al. 2008). For example, in whales, pinnipedes, and sirenia, unihemispheric slow-wave sleep (USWS) is exhibited, in which only one hemisphere of the brain is sleeping. It is proposed that within the specific ecological framework of the *Astyanax* cave fish, reduced sleep is no loss but may instead be an adaptation to an environment, where the change between night and day has lost its biological function. It is as yet unknown whether the reduced sleep duration might be compensated for by the physiological intensity of sleep in the cave fish. Thus, sleep in *Astyanax* is an independent peripheral outcome of circadian rhythmicity, which changed constructively under altering environments in the cave fish.

Taken altogether, the reduced sleep pattern might be a constructive rather than a regressive trait. The cave fish have evolved a dramatically reduced sleep phenotype to increase the time available for foraging (Duboué et al. 2012), which would be a gain and no loss. Resting and activity patterns have been uncoupled from circadian rhythm by destructive mutations, because they have lost their biological function in constant darkness and are no longer submitted to selection.

## 6.11 Egg Yolk Content

Few cave species like the cusk eels of the genus *Typhliasina* and *Lucifuga* (Bythitidae) or the live-bearing toothcarp *Poecilia mexicana* (Poeciliidae) produce live offspring. They do so in rather low numbers of 2–15 juveniles. However, most cave fish lay large numbers of eggs (up to about 1000) in every spawning process. At the beginning, the fish embryos exclusively subsist on yolk. Comparing the amount of egg yolk content it was found that between the troglophile surface catfish *Rhamdia laticauda* and its cave sister species *R. zongolicensis* and *R. reddelli* no differences exist (Wilkens 2001). In contrast, *Astyanax* cave fish provide an example of the yolk content being significantly increased (by about 50%) in comparison to the surface sister form. This was shown by the study of five different cave populations (Micos, Chica, Pachón, Piedras, Yerbaniz) (Figs. 6.11 and 6.12). The chemical composition of yolk (e.g. fat, protein) is nearly identical between surface and cave forms.

Due to the higher quantity of yolk reserves in the egg, the *Astyanax* cave fish juveniles grow for a longer time and are larger than the surface fish (Hüppop and Wilkens 1991) when they start external feeding. For example, Pachón cave fish larvae do so at an age of 4 days after fertilization, compared with 3 days in the surface fish. Because of their size they are probably able to catch slightly larger prey (Fig. 6.12) (Hinaux et al. 2011; Hüppop 1987, 1989; Yamamoto et al. 2009).

Mean and distribution of yolk content in the F1 crossings show a more or less intermediate position between surface and phylogenetically old Pachon or Piedras

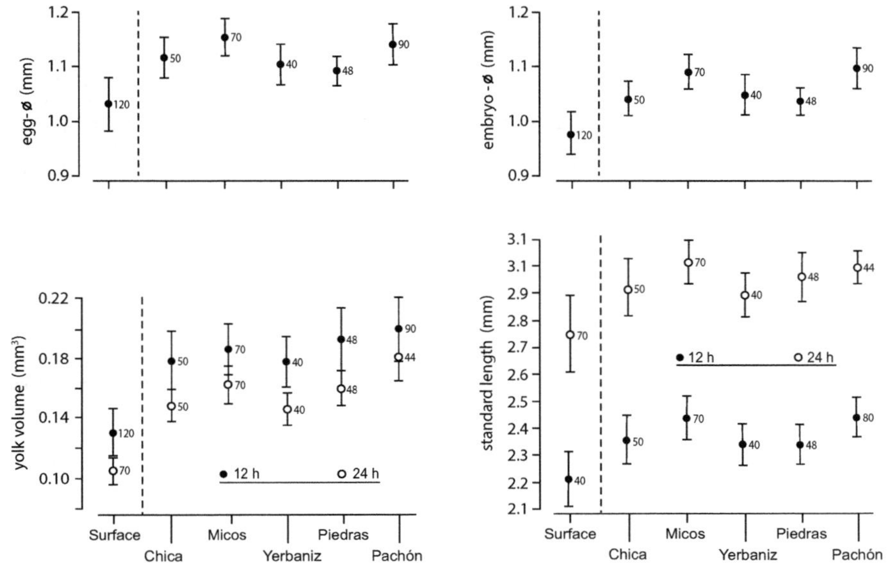


**Fig. 6.11** Means and standard deviation of egg and embryo size of *Astyanax* surface fish and five cave fish populations at 12 (*closed circle*) and 24 h (*open circle*) after fertilization, numbers give sample size (adapted from Hüppop and Wilkens 1991)

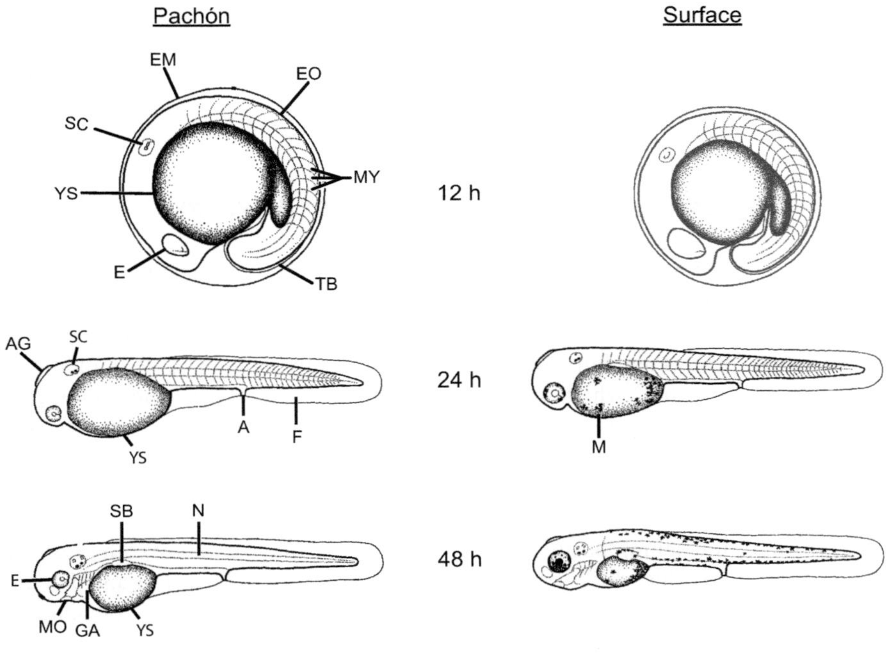


**Fig. 6.12** Comparison of yolk content and growth in *Astyanax* surface and Pachón cave fish at 12, 24, and 48 h after fertilization. *A* anus, *AG* adhesive gland, *E* eye, *EM* egg membrane, *EO* embryo, *F* fin fold, *M* melanophore, *MY* myomeres, *N* notochord, *NP* olfactory pit, *SB* swim bladder, *SC* statocyste, *TB* tail bud, *YS* yolk sac (adapted from Hüppop and Wilkens 1991)

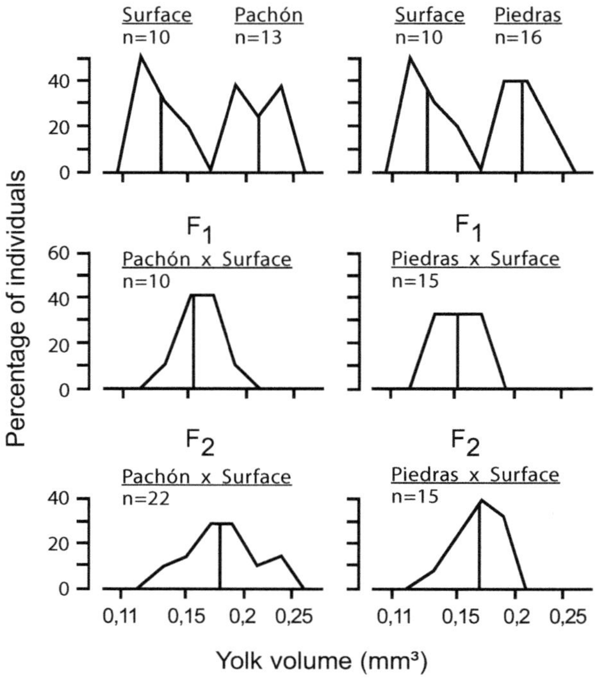


**Fig. 6.13** Yolk content of *Astyanax* surface, the Pachón and Piedras cave fish as well as in their F1 and F2 crossings (adapted from Hüppop and Wilkens 1991)

cave fish. These findings insinuate that the increase of yolk content is polygenic, showing epistatic gene effect (see Sect. 6.23) (Fig. 6.13) (Hüppop and Wilkens 1991).

## 6.12 Aggressive Behavioural Patterns in *Astyanax*

Animals competing for food and mates display species-specific aggressive behaviours independently of whether they live in dark caves or in lit environments. Visual signals often play an important role in aggressive behaviour, but cannot be used in darkness (Parzefall 2000). Caves are mostly colonized by troglophile species such as catfish, which are already preadapted to darkness and do not alter their aggressive behaviour in the new environment. However, when such preadaptations are lacking, changes in aggressive behaviour occur, originally visually triggered traits get lost and new ones are developed (Parzefall and Trajano 2010).

*Astyanax* surface fish exhibit a visually released complex aggressive behaviour consisting of threatening postures and attacks, which may be followed by fights if the attacked individual does not take flight (Parzefall 1983; Parzefall and Trajano 2010). The threatening surface fish in general enlarges its body size by spreading its fins (aggressive fin spreading) and snake swimming (Fig. 6.14a, b). In escalated fights, tail beating (Fig. 6.14c), circling (Fig. 6.14d), ramming attempts, ramming,

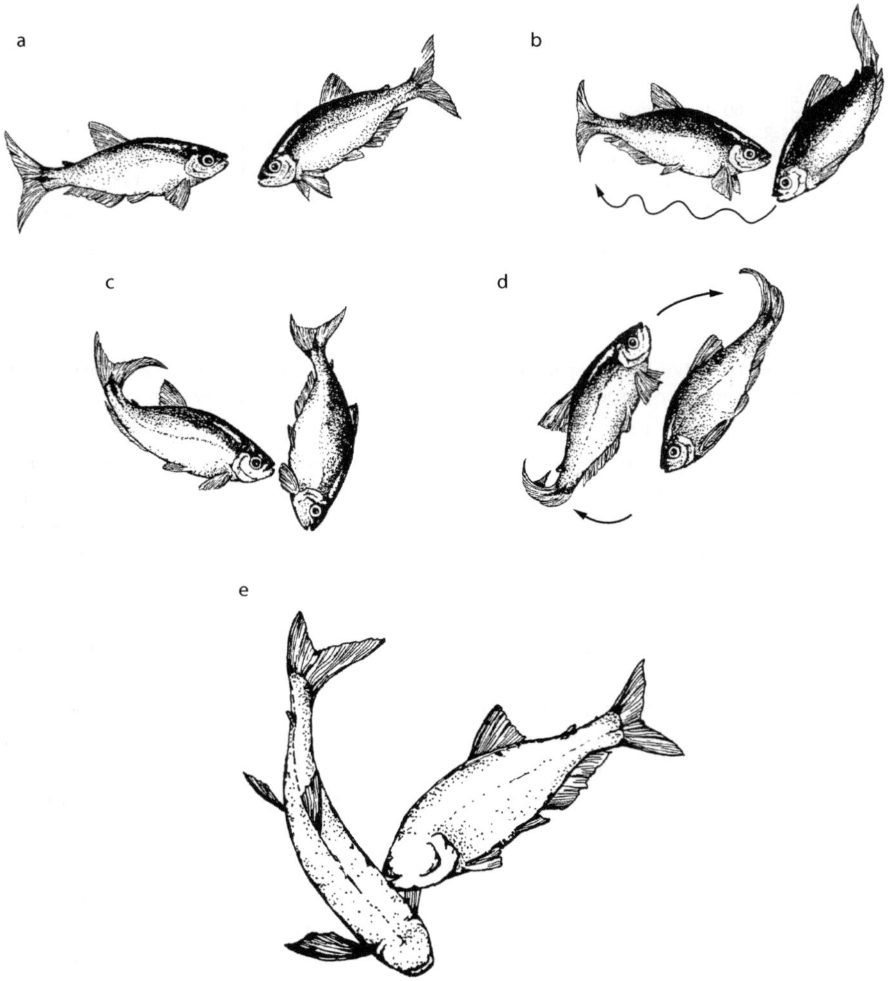


**Fig. 6.14** Aggressive patterns in *Astyanax* surface fish. (**a**) Fin erection: the head-down position of the right fish expresses a higher aggressive motivation, (**b**) snake swimming of the right fish and aggressive fin erection by the left one, (**c**) ramming attack of the left fish, (**d**) circling and tail beating, (**e**) ramming and circling in the Pachón cave fish (adapted from Parzefall and Hausberg 2001)

and biting is performed by the attacking individual (Parzefall and Hausberg 2001). The intensities of the aggressive patterns of fin spreading, ramming attempts, ramming and tail beating are positively correlated (Fricke and Parzefall 1989).

In contrast to the surface fish, strongly eye-reduced cave populations like the SEP Pachón, Yerbaniz, and Piedras cave fish are no longer able to perceive and exhibit visually released aggressive behaviour like fin spreading and snake swimming. Instead, other elements of surface aggressive behaviour like biting are

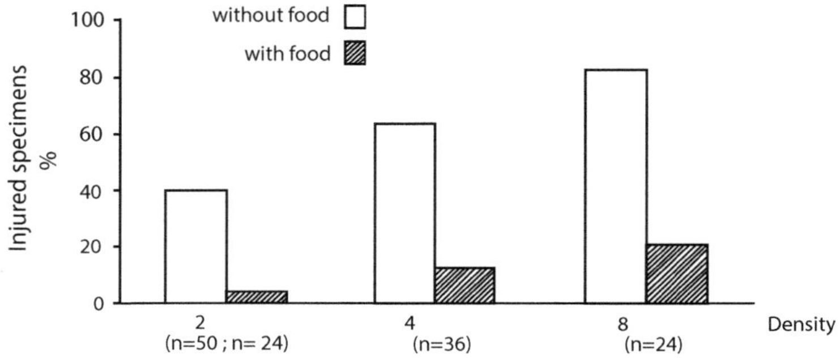


**Fig. 6.15** Percentage of injured specimens of the Pachón cave fish in dependence of food supply and density. $n$ = number of specimens studied (adapted from Hausberg 1995)

exhibited (Fig. 6.14e). As close proximity and direct body contact were observed to be the premise of such aggressive actions in the SEP cave fish, it is assumed that a different sense like the lateral line is involved in triggering it. As a consequence of these fights the cave fish also get hurt, shown by a loss of scales and parts of the fins, especially of the caudal one (Hausberg 1995). The amount of injuries depends on the population density (Fig. 6.15).

Aggressive behaviour in surface and cave *Astyanax* is a density controlling, sex-independent mechanism (Parzefall 1983). In the surface fish, escalated fighting usually does not take place as long as the fish have enough space while schooling in their natural habitats like rivers, cenotes, and lakes. The threatening behaviour of fin spreading and snake swimming will suffice to avoid injuring fights. However, during the dry season when food gets scarce and the surface fish get separated in shrinking pools, small feeding territories are defended (Parzefall and Trajano 2010). When there is no chance to escape or hide, fights occur leading to severe injuries, and the weaker opponent may even get killed. In the cave fish populations, territories are only built and defended under food deprivation (Hausberg 1995).

In absolute darkness, neither juvenile nor adult surface fish display aggressive behaviour or establish territories at all (Fig. 6.16). The finding that surface specimens do not show scale loss or fin injuries supports the theory that no fights take place under such conditions (Hausberg 1995; Parzefall 1983, 2000; Wilkens 1988). Therefore, it is proposed that the aggressive behaviour of the *Astyanax* surface fish is exclusively visually released.

The optical releasers for aggression were studied by observing the choice-preference behaviour of single specimens with respect to two simultaneously presented signals (Langecker et al. 1995). Three experimental tanks were arranged in a linear order. The experimental set-up ensured that only optical signals could be transmitted. The central tank was marked at its centre by a line dividing it into left and right compartments. Two alternative signals were presented at the left and the right side of the central tank. As signals, live surface and cave fish as well as surface dummies were tested. Among the observable agressive behavioural patterns, only

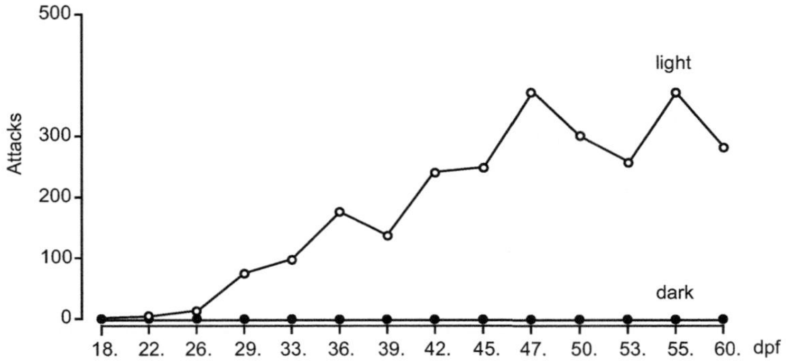


**Fig. 6.16** Ontogeny of aggressive behaviour of *Astyanax* surface fish at light and in absolute darkness (number of attacks/12 min). *dpf* days post-fertilization (adapted from Hausberg 1995)

"ramming attempt" performed at the left or the right side of the central tank was counted as a general measure for aggression because of its high frequency and distinctness.

All of the 20 surface individuals tested attacked the moving, naturally coloured live surface signal fish rather than the alternatively presented live Pachón cave fish (Fig. 6.17a). A significantly lesser number of tested surface specimens (79%) preferred the moving, naturally coloured 3D surface dummy when presented alternatively with a stationary surface dummy (Fig. 6.17b). This shows that both "natural shape and colouration" as well as "locomotion" are effective releasers of aggression, but that the combination of both is important in the surface fish. Movement alone is less effective. These findings are also stated when testing the signal combination of a live moving Pachón cave fish presented as an alternative to a stationary surface dummy (Fig. 6.17c) as well as a moving 2D square plastic sheet presented as an alternative to a stationary dummy (Fig. 6.17d). In both experiments the moving object was preferred by a very low number of tested surface specimens, however.

Regression of the releaser "natural shape and colouration" has obviously occurred in the cave fish. It can be assumed that this ability of perceiving visual releasers got reduced because it is no longer controlled by stabilizing selection. Whereas all surface specimens tested attacked the surface signal fish when alternatively presented with the Pachon signal fish, a significantly lower number of the F1 hybrids (surface × SEP Pachón) (65%) (Fig. 6.17a) as well as of the VEP Micos fish (72%) (Fig. 6.17a) attacked the live surface signal fish. From the nearly intermediate number of attacks in the F1 it can be concluded that the releaser is completely reduced in the SEP Pachón cave fish and from the lower number found in the VEP Micos fish that it is on the way to reduction.

The choice experiments also revealed a change of the releaser "locomotion" in cave specimens. Whereas only some of the surface specimens (22%) attacked a moving object that had no natural shape but consisted just of a square piece of plastic (Fig. 6.17d), higher numbers of the tested F1 hybrids (65%) and of the Micos

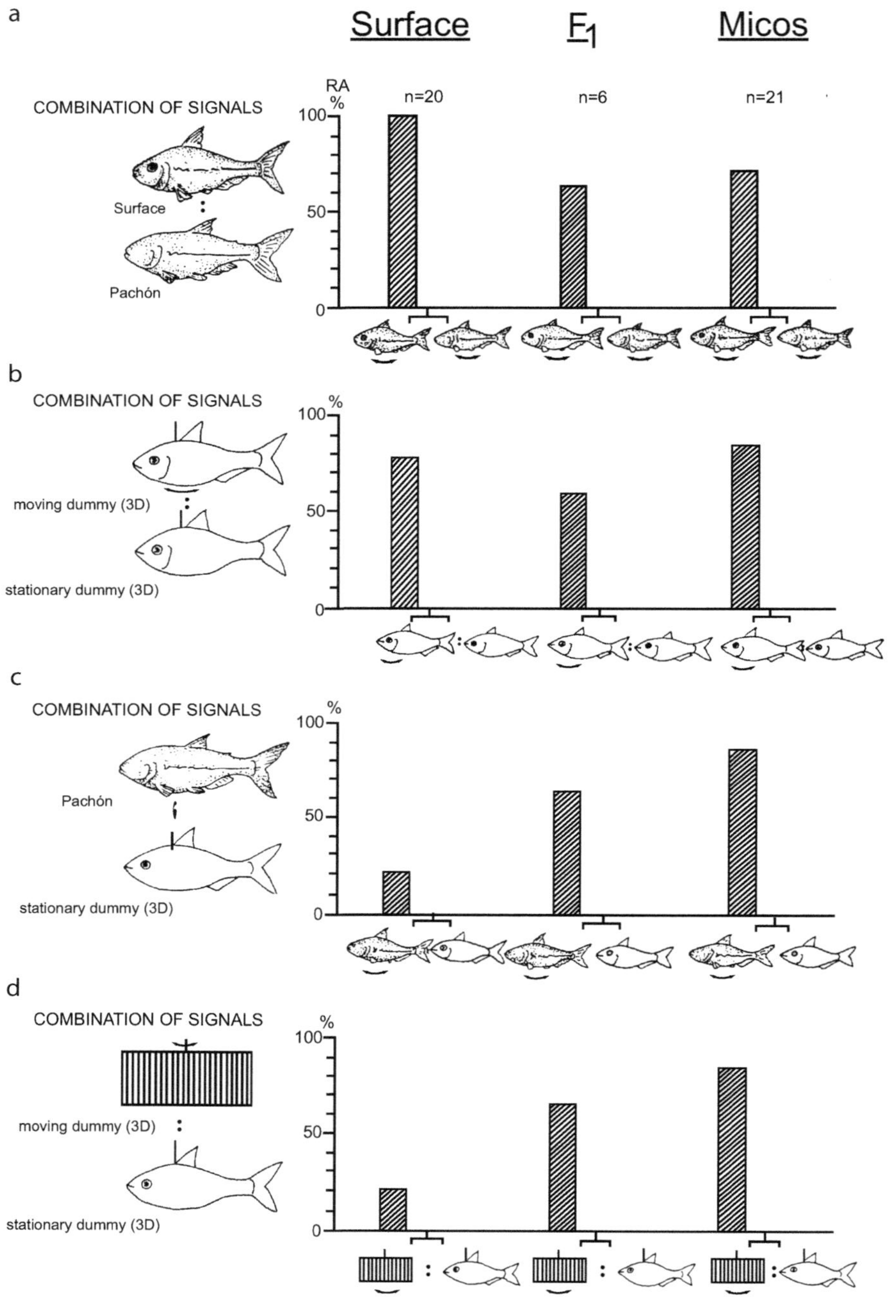


**Fig. 6.17** Percentage of surface, F1 crossing (SEP Pachón cave × surface fish), and VEP Micos cave fish showing a significant preference for aggressive ramming attempts to distinct combinations of signals. *n* Total number of specimens tested, *RA* ramming attempts, *SEP* strongly eye- and pigment-reduced, *VEP* variably eye- and pigment-reduced (adapted from Langecker et al. 1995)

fishes (85%) preferred to attack all moving objects regardless of their shape (Fig. 6.17c, d). From this is concluded that the ability of perceiving the releaser "locomotion" has been improved in the cave fish.

Thus, in all cave fish populations the releaser has changed, because aggression is almost exclusively triggered by visual stimuli in the surface ancestor. The stimulus "locomotion" is probably perceived by the lateral line organ and has come to play a new role in releasing the aggressive behaviour of *Astyanax* cave fish. Whereas it was found that in the SEP Pachón and Piedras cave fish the visually released trigger of aggressive behaviour is reduced and has been completely replaced by the trigger "locomotion", the VEP Micos cave fish has an intermediate position between surface and SEP cave fish. Testing of well sighted Micos cave fish revealed differing degrees of intensity of the visually triggered aggressive attacks performed by them. In some specimens they are even lacking (Hoffman and Hausberg 1993). In the mean, however, the intensity of the visually triggered aggressive behaviour is lower than in the surface fish. Like in eyes and pigmentation, the visually triggered aggressive behaviour is lower in the VEP Micos cave fish and comprises an intermediate stage between the SEP and surface fish.

The number of aggressive reactions measured as ramming attacks in the F1 as well as the F2 crossings between the surface and the SEP cave fish indicates that the reduction of the visually triggered aggressive behavioural traits in the SEP cave fish relies on a genetic basis. The mean number of attacks is lower and they become completely manifested only at an ontogenetically later stage compared with the surface fish (Fig. 6.18) (Hausberg 1995; Langecker et al. 1995). As all Micos specimens tested had large eye size and good vision ability, but showed differing degrees of visually triggered aggressive behaviour, it was concluded that eye size and aggressive behaviour inherit independently (Hoffman and Hausberg 1993; Langecker et al. 1995; Parzefall and Hausberg 2001).

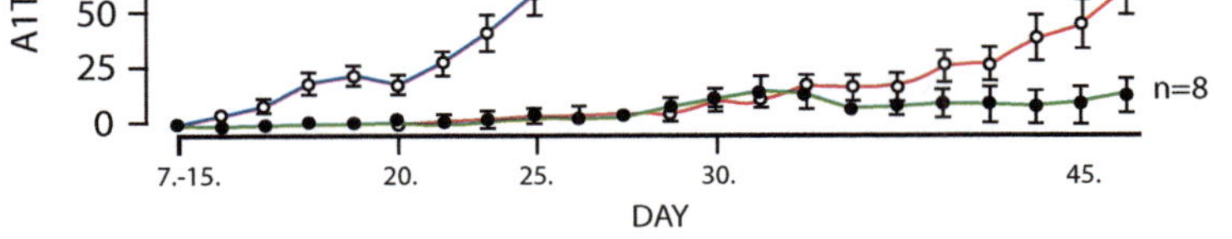


**Fig. 6.18** Ontogeny of the attack rate of *Astyanax* surface fish and F1- and F2-crossing hybrids with the Pachón cave fish (mean and standard deviation, adapted from Parzefall and Hausberg 2001)

In several studies it was claimed that surface *Astyanax* would be aggressive in darkness. For example, Burchards et al. (1985) observed that surface fish show aggressive behaviour in darkness. It was later found that the experiments were not conducted in complete darkness and that the light from the control lamp for the heat rod and the infrared devices applied provided enough illumination for the surface fish to build and defend territories (Hausberg 1995; Burchards, personal communication). Aggressive behaviour was also observed in lens-extirpated surface fish, because it was assumed that they would mimic reduced cave fish eyes not able to see (Espinasa et al. 2005). However, it was not considered that these eye residuals still contain intact visual cells (see Fig. 1D in Espinasa et al. 2005) and operate like a pinhole camera. They may perceive moving objects at least as diffuse images, which in *Astyanax* surface fish suffices to release aggressive reactions. These findings rely on inappropriate experimental conditions.

It was furthermore claimed that the SEP cave fish had lost the aggressive behaviour and hypothesized that during their evolution in darkness they had made an "evolutionary shift from fighting to foraging" under the bias of the cave fish being "food addicts" (Elipot et al. 2013). These results may have come about by applying a too-short acclimatization time of only few hours, not considering that adult cave fish start performing aggressive behaviour only after 2 days (Hausberg 1995). Besides this, aggressive behaviour in the crossings between surface and SEP cave fish starts at later ontogenetic stages (Fig. 6.18) and it is as yet unknown whether it is the same in the cave fish. Therefore, it cannot be excluded that the alleged foraging behaviour observed in these experiments performed with juvenile cave fish are only random clashes caused by swimming activity performed by the cave fish to get spatial orientation after having been introduced into a new environment (see Sect. 6.5).

## 6.13 Dorsal Light Reaction

The dorsal light reaction enables fish to orientate the body in a vertical position according to two factors, namely the light input from above and earth gravity from below. This behaviour furthermore serves to camouflage the fish against aerial predators by adapting the dorsal melanophore and iridophore colouration to that of the albedo of the underground. Simultaneously, it provides camouflage against aquatic predators attacking from below, because the silvery shine caused by guanophore colour cells on the ventral side clear away the body contours of the fish at the water surface.

When kept in darkness, both surface and cave fish swim in an upright position, which shows that solely one of the two factors, namely earth gravity, is sufficient for an upright orientation. However, when a light beam is applied in the dark at a 90° angle from the side, the surface fish will incline its body to a 45° angle. Because of the missing eyes, blind SEP cave fish like Pachón or Piedras show no reaction (Fig. 6.19). The eyed F1-crossing hybrids between both forms manifest an intermediate tilt under these experimental conditions. In F2 hybrids equipped with good vision, the tilt angles range between nearly missing and the mean exhibited by the

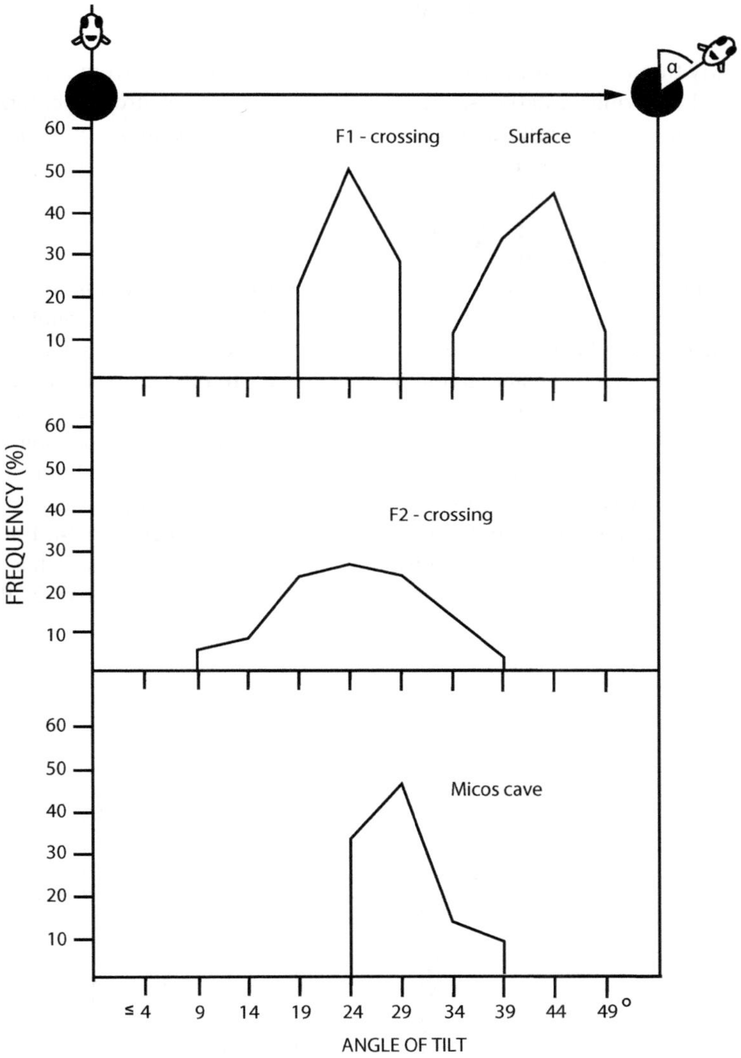


**Fig. 6.19** Frequency distribution of the angle of tilt of *Astyanax* surface fish, the F1- and F2-crossing hybrids with the SEP Pachón, and the VEP Micos cave fish, when illuminated at an angle of 90° from the side. *SEP* strongly eye- and pigment-reduced, *VEP* variably eye- and pigment-reduced (adapted from Langecker 2000)

surface form. These results suggest that part of the dorsal light reaction is strongly reduced in the SEP cave fish and only the earth gravity part is still functional. The VEP Micos cave fish lie between the mean angle of the F1 crossing between the surface and SEP cave fish, and the surface fish. The intermediate distribution of the angles in the F2 and in the F1 crossing between surface and SEP cave fish shows that the dorsal light reaction relies on a polygenic basis (Langecker 1993).

## 6.14 Phototactic Behaviour

The mostly troglophile surface ancestors of cave species derive from forms already preadapted to living in darkness. Avoiding light they show strong photonegative behaviour and mostly prefer a nocturnal way of life. For example, photonegativity was found in the Somalian cave fish *Phreatichthys andruzzi* (Cyprinidae) and *Uetgitglanis zammaroni* (Claridae) (Ercolini and Berti 1975). In *P. andruzzi* it was revealed that encephalic photoreception could explain the observed behavioural spectral sensitivity (Tartellin et al. 2012).

It is also described for the Mexican surface catfish *Rhamdia laticauda* (Hepapteridae) (Langecker 1992a; Wilkens 2001). Therefore, photonegative behaviour has long been looked upon as a prerequisite of surface species for cave colonization as well as an adaptation of cave species to prevent them from leaving the dark environment. However, the study of two cave sister species of surface *R. laticauda*, *R. zongolicensis* and *R. reddelli*, revealed that, in comparison with the surface species, photonegativity is less strong in them and shows high interindividual variability (Fig. 6.20), although photosensitivity is most probably still functional because of the pineal containing visual cells (Eilertsen et al. 2014; Langecker and Wilkens 1992). This could imply that the phototactic behaviour has lost its biological function in these cave species and is regressive, because in the vast

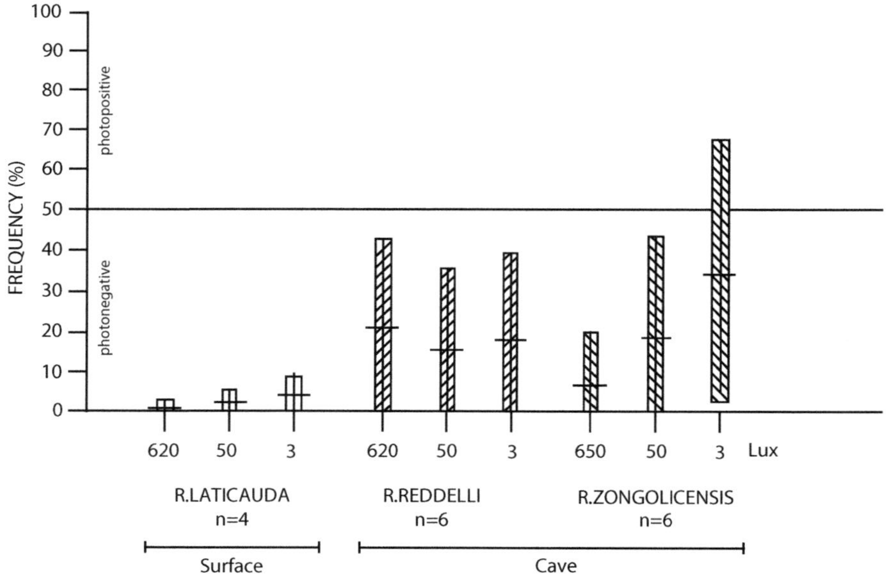


**Fig. 6.20** Phototactic reactions of the surface catfish *Rhamdia laticauda* and its cave sister species *R. reddelli* and *R. zongolicensis* at different light intensities (650, 50, 3 lux). Shown are mean response, standard deviation (*vertical line within box*) and 95% confidence limits of the mean (*box*) (adapted from Langecker 1992a)

underground cave system the fish rarely, if at all, make contact with the influence of surface light.

In *Astyanax* surface fish it has been shown that phototactic behaviour changes during life history and therefore has to be evaluated in its ecological context (Langecker 2000). Whereas juveniles prefer lighter zones, even reacting photo-positively, adult specimens are photonegative in bright light but indifferent in low light intensities. This differing behaviour may favour the utilization of different foraging areas by both groups as well as save the juveniles from being preyed on or attacked by the larger adult fish (Parzefall 1983), which prefer to hover in lesser illuminated places (Langecker 2000; Romero 1985) (Fig. 6.21).

For a comparison of phototactic behaviour between adult *Astyanax* surface and cave fish, the visual orientation was experimentally excluded by the complete enucleation of the eye ball in the surface fish. Under these conditions, the intensity of the photonegative reaction of the adult surface fish increases when compared with untreated surface fish and—except for low light intensity—is much stronger than in the cave fish. After additional ectomy of the pineal organ, the photonegative reaction in the surface fish is nearly neutral and the same as in normal, untreated cave fish (Fig. 6.22) (Langecker 1992a, b). The slightly negative phototactic behaviour observed in pineal-ectomized cave specimens possibly relies on the existence of melanopsin-containing intrinsically photosensitive retinal ganglion cells (ipRGCs) (Lucas et al. 2014).

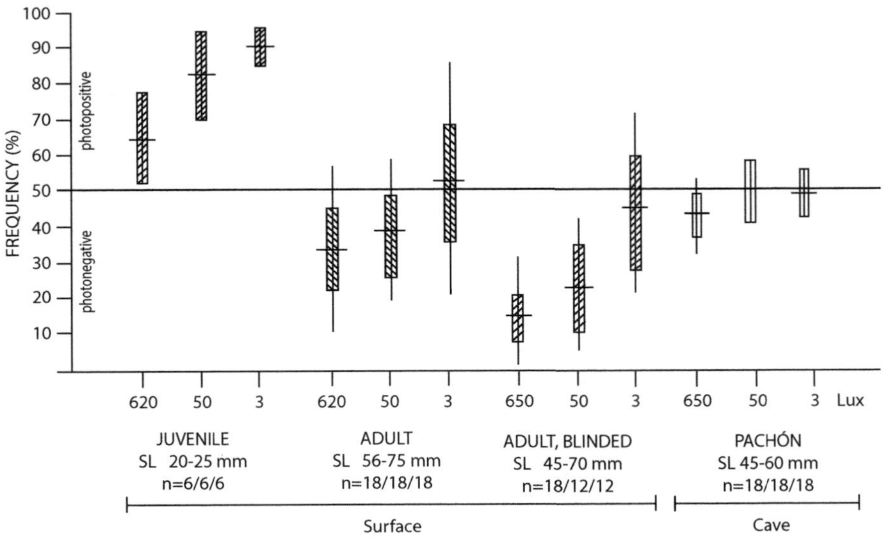


**Fig. 6.21** Photonegative behaviour of juvenile eyed and blinded (eye enucleated) adult *Astyanax* surface fish and of adult Pachón cave fish. For explanation see Fig. 6.20 (adapted from Langecker 1992a, 2000)

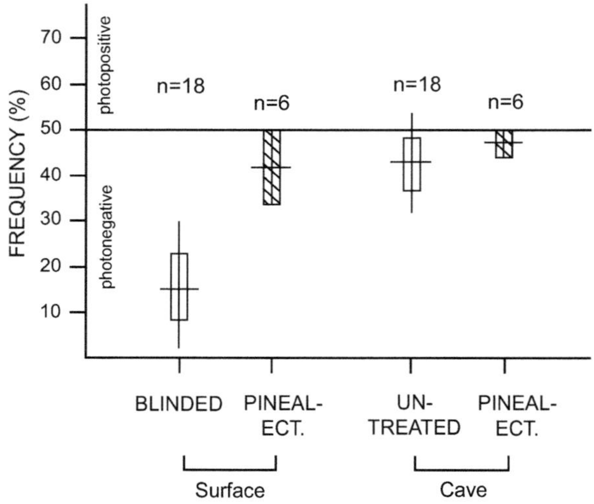


**Fig. 6.22** Influence of the pineal organ on the phototactic behaviour in solely eye enucleated and eye enucleated as well as pineal ectomized *Astyanax* surface fish and in solely pineal ectomized Pachón cave fish. For explanation see Fig. 6.20 (adapted from Langecker 1992a)

It is suggested that the generally lower degree of photonegativity in the Pachón cave fish compared with the surface fish is not due to morphological degeneration of the light-perceiving structures, but is caused by the regression of the photonegative behaviour (Langecker 1992a).

This was also corroborated by studies of the phylogenetically old Tinaja cave fish, which showed no preference for the dark or the light (Kowalko et al. 2013b).

In crossings between surface and SEP cave fish, it was revealed that the F1 hybrids display strong negative phototactic behaviour comparable to the surface fish whereas the distribution curve in the F2 crossing is bimodal (Kowalko et al. 2013b). From this it can be interpreted that a polygenic system exhibiting epistatic gene effect is responsible for the phototactic behaviour (see Sect. 6.23).

## 6.15 Pineal Organ

The pineal organ is a small endocrine gland in the vertebrate brain that is photoreceptive because of the presence of exorhodopsin and melanopsin in photoreceptor cells structurally resembling the cones of the lateral eyes (Fig. 6.23) (Eilertsen et al. 2014). They regulate the melatonin synthesis important for photoentrainment of the circadian rhythm directly (Falcón et al. 2007, 2010). The nonvisual opsin exorhodopsin, which has been shown to be a pineal-specific opsin expressed from early in development of teleosts and has been indicated to have a role in the circadian rhythm, was evidenced in the *Astyanax* cave fish (Parry et al. 2003; Pierce et al. 2008).

Comparative histological studies of the pineal of adult *Astyanax* surface fish, the SEP Pachón and the VEP Chica cave fish revealed only minor differences between each other (Herwig 1976; Langecker 1992a). In the Chica and the Pachón cave

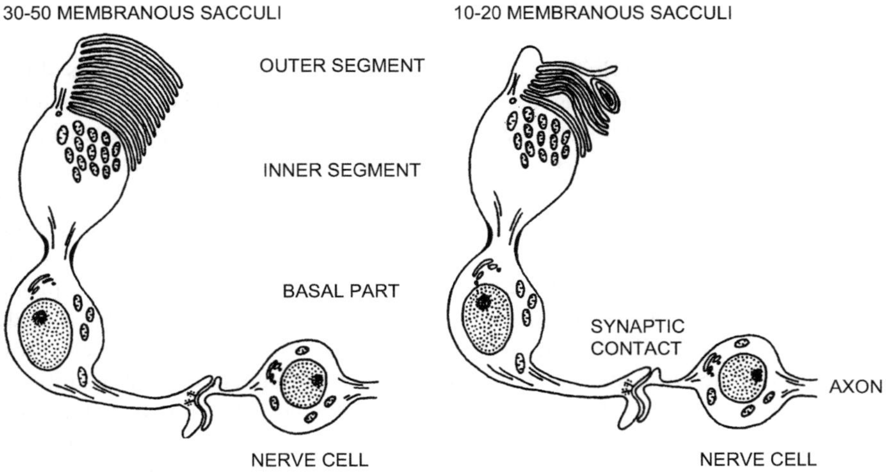


**Fig. 6.23** Scheme of a single pineal photoreceptor cell of the *Astyanax* surface fish (*left*) and the Pachón cave fish (*right*) (adapted from Langecker 1992a, b)

populations, it was found that the size of the end vesicle was diminished when compared with the surface fish. Furthermore, the visual cells' outer segments have a reduced number of discs and are characterized by a large structural variability containing well differentiated as well as disorganized outer segments (Fig. 6.23) (Herwig 1976; Langecker 1992a). Also, in the blind cave salamander *Proteus anguinus* and the cave catfish *Rhamdia zongolicensis* such regressive processes restricted to the outer segments were observed whereas the surface sister species *R. laticauda* shows a well developed pineal organ (Langecker and Wilkens 1992; Kos and Bulog 2000). Study of the Chica and Pachón cave fish revealed that the outer segments of the pineal seem to be submitted to ontogenetic regression (Herwig 1976; Langecker 1992a). Thus, like the visual cells of the cave fish eye, those of the pineal organ seem to be submitted to a process of ontogenetic regression, but which does not end in total reduction. Compared with other light-dependent morphological structures, the degree of morphological reduction of the pineal organ as studied in the Pachón and the Chica cave fish is low, which can be attributed to the extant secretory activity and its function in regulating circadian rhythmicity (see Sect. 6.16).

Photosensitivity of the pineal was also confirmed by studies of the phototactic behaviour of adult cave and surface fish (see Sect. 6.14) (Langecker 1992a). Photoreception has already been detected in larval fish. *Astyanax* surface and cave fish larvae swim upward to the water surface when shaded from above, exhibiting the so-called shadow response. It is strongest at 1.5 days post-fertilization and has almost vanished at 6.5 days (Yoshizawa and Jeffery 2008).

## 6.16 Circadian Rhythm

The control of circadian rhythms and the coordination of different oscillators of fish are regulated by the oscillating melatonin production in the pineal organ. Melatonin as a common output signal of the vertebrate circadian clock is produced in the pineal photoreceptors and also in the retina. Different from retinal melatonin, pineal melatonin is released into the blood stream and the cerebrospinal fluid as soon as it is synthesized. However, the pineal organ is considered to serve as the central pacemaker in fish. Here the circadian clock drives rhythmic synthesis of the hormone melatonin. Studies in the pike (*Esox lucius*) have shown that each pineal photoreceptor cell contains a circadian mechanism, so that one pineal organ is made of multioscillatory cellular units (Bolliet et al. 1996). The rhythms of gene expression, enzyme activity, and melatonin release free run under constant conditions. Melatonin levels are high at night and low during the day. Photoreceptor cells, obtained from dissociated pineal organs and cultured either alone or together, maintain a rhythmic secretion of melatonin under alternating light and darkness (LD) and continual darkness (DD) for up to a week (Falcón 1999; Falcón et al. 2007, 2010; Foulkes et al. 2016; Idda et al. 2012).

In *Astyanax*, radioimmunassays of extirpated pineals in superfusion experiments were performed in the surface fish, the SEP Piedras, and the VEP Micos cave fish. These studies revealed a well detectable melatonin production. Rhythmic modulation can still be observed, indicating its role in endogenous circadian rhythmicity (Missal 1994; Wilkens et al. 1993) (Fig. 6.24). In surface and cave fish, the melatonin production was found to be high during the subjective night phase, lowered at the beginning of the subjective day phase, and rose again at the subjective night phase. After entrainment in LD 12:12 h the melanin production oscillates free running in DD for several days. The three *Astyanax* populations differ concerning the secreted amount of melatonin as well as concerning the secretional profile. The period length ($\tau$) was 23.8 h in the surface fish, 26.4 h in the Piedras cave fish, and 24.2 h in the Micos cave fish. After light stimulation, melatonin production in all pineals is immediately lowered (Fig. 6.25). This can be attributed to regulated transcription and stability of serotonin-*N*-acetyl-tranferase (AANAT), a key enzyme in melatonin synthesis (Ziv et al. 2005), still functioning.

The melatonin levels in the pineals of the three *Astyanax* populations differed significantly (Wilkens et al. 1993). The pineal end vesicle of surface *Astyanax* contained on average 6.3 pg melatonin during the light and 119.5 pg during the dark period in February at LD 12:12 and did not significantly change at differing day length in June at LD 18:6 (Fig. 6.26). In contrast, the melatonin content of the SEP Piedras cave fish was high during daytime and not significantly different at night in February. In June, however, the daytime level was much lower, although still higher than in the surface fish. This could be caused by the light perception of the cave fish not being equally efficient as that of the surface fish, because of eye loss and the position of the light sensitive pineal in the interior of the head. Further studies should reveal whether, under permanent darkness, the SEP Piedras cave fish might exhibit more or less constant melatonin levels.

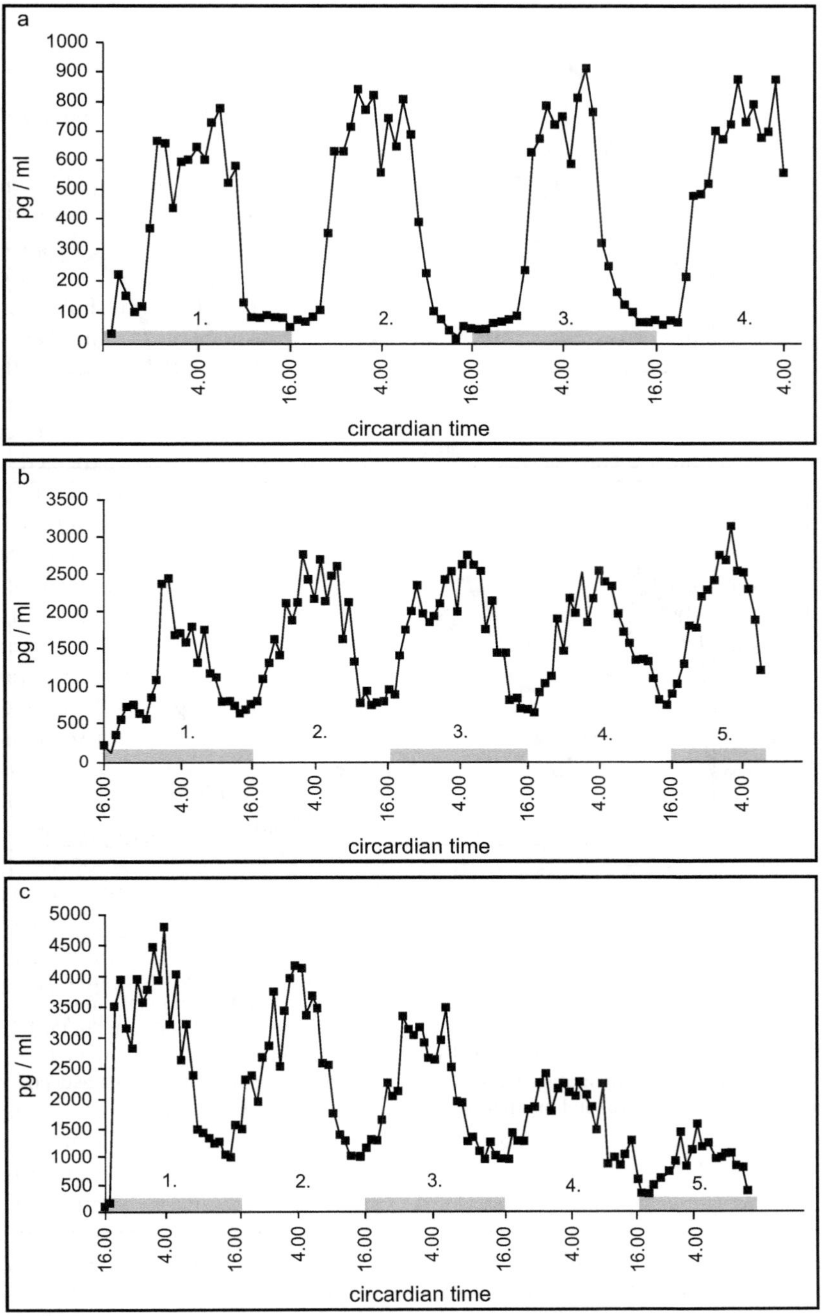


**Fig. 6.24** Melatonin production of an isolated individual pineal organ of an *Astyanax* surface fish (**a**), the VEP Micos cave fish (**b**), and the SEP Piedras cave fish (**c**) all kept in DD after LD 12:12.

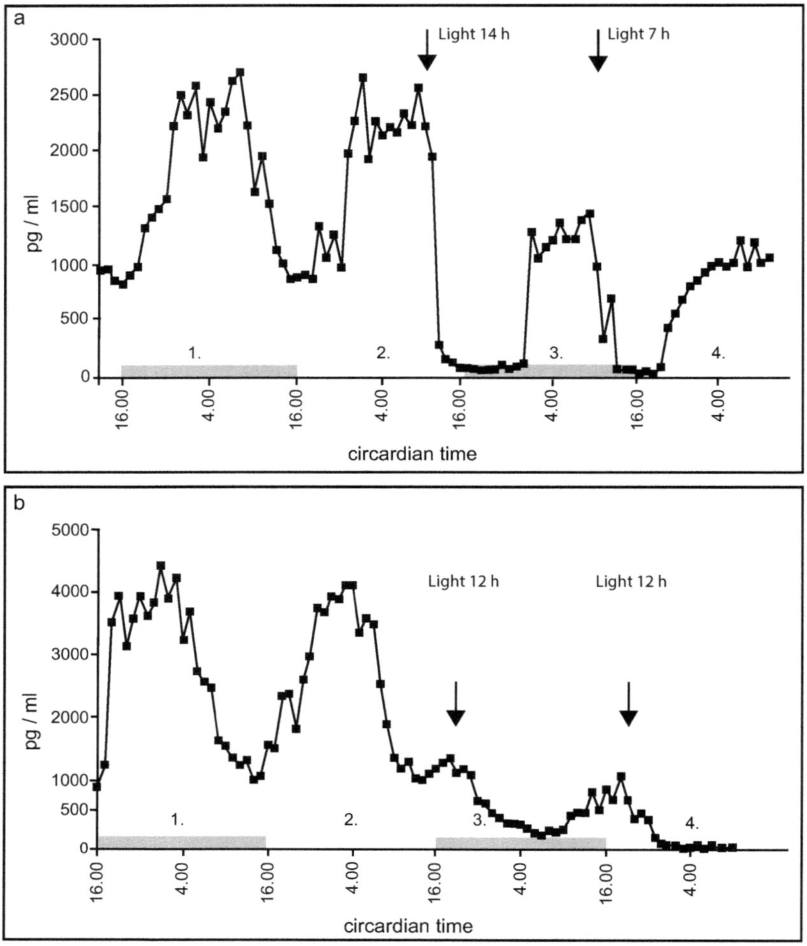


**Fig. 6.25** Influence of light stimulation on melatonin production of an isolated individual pineal organ of the VEP Micos cave fish (**a**) and the SEP Piedras cave fish (**b**) kept in DD after LD 12:12. Light stimulation is procyclic in Micos for 14 and 7 h, respectively, and anticyclic in Piedras for 12 h. Days were measured in Micos from 1 pm to 1 am and in Piedras from 4 pm to 4 am. *DD* continual darkness, *LD* alternating periods of light and dark, *SEP* strongly eye- and pigment-reduced, *VEP* variably eye- and pigment-reduced (adapted from Missal 1994)

---

**Fig. 6.24** (continued) Analysis was by superfusion and radioimmunoassay, probes were taken every hour for 4 days (surface fish) and 5 days (cave fish), days were measured from 4 pm to 4 am. *DD* continual darkness, *LD* alternating periods of light and dark, *SEP* strongly eye- and pigment-reduced, *VEP* variably eye- and pigment-reduced (adapted from Missal 1994)

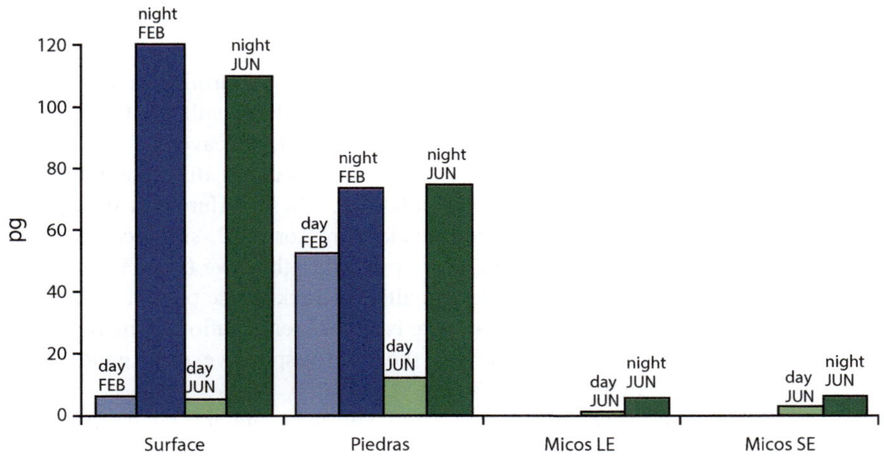


**Fig. 6.26** Circadian changes of the mean melanin content in the pineal end vesicle of the *Astyanax* surface, the SEP Piedras, and the VEP Micos cave fish in February (LD 12:12) and June (LD 18:6). Micos LE and SE indicate large and small eyed specimens, respectively. *LD* alternating periods of light and dark, *SEP* strongly eye- and pigment-reduced, *VEP* variably eye- and pigment-reduced (adapted from Wilkens et al. 1993)

In these studies, the melatonin content in the pineal of the VEP Micos cave fish diverged even more from the surface fish than the SEP Piedras cave fish. Regardless of the variable eye size of the Micos specimens analyzed, the day-time value was minute in February and June (1.5 pg/3.1 pg) and the night-time value for the respective seasons (5.8 pg/6.6 pg) was within the range of the low day-time values of surface fish. These unexpected discrepancies of and between the SEP Piedras and the VEP Micos cave fish remain to be studied. However, altogether the results have shown that the cave fish melatonin level has changed in both cave populations.

Rhythmicity between day and night time was conserved in all experiments. The *Astyanax* surface and cave fish pineals were found to still be oscillating systems with endogenous circadian rhythmicity, which may be influenced by external factors like light stimuli. This extant functionality might explain why, in contrast to the cave fish retina, the pineal photoreceptors show almost no regression (see Sect. 6.15). As all fish from the SEP Piedras and the VEP Micos populations are able to exhibit circadian melatonin production, it is suggested that this feature is conserved in the cave fish. Further research will need to reveal, however, whether the nearly equal content of melatonin found in the SEP Piedras during the conditions of poor day light in February correlates with altered behavioural activities like loss of sleep as observed in the SEP Pachón cave fish (see Sect. 6.10) (Fig. 6.24). Possibly it is even a prerequisite for breeding success in the cave fish (see Sect. 6.2).

Molecular studies have revealed that circadian rhythms in the clock gene period 1 (*per1*), which represents a key element of the core clock mechanism and shows high amplitude circadian oscillations in most teleosts, are retained in the Pachón and Chica cave fish but exhibit substantial alterations (Fig. 6.27) (Beale et al. 2013). In LD, *per1* levels oscillate in the two cave populations, confirming that the ability to detect light and generate molecular circadian oscillations still exists. However, there are differences between cave and surface fish. First, the cave fish *per1* rhythm is lower in amplitude and, second, the entrained phase is different with peak expression 6 hours later in both cave populations. These differences then persist in DD (Fig. 6.28). In the light-induced clock repressor *per2*, significantly raised levels compared with surface fish in DD are present in the cave fish. This might be responsible for the reduced amplitude and altered clock phase of molecular clock rhythms in the cave fish. It appears that there is a clear perturbation of the light input pathway in the *Astyanax* cave fish with the light-responsive *per2* gene constitutively expressed at high levels.

In the Chica cave fish, expression levels of *per1* under DD do not show rhythmicity and are continuously lower in the wild than in Chica fish kept in the laboratory as well as in surface fish, whereas *per2* is raised in the field comparable to the laboratory. Thus, no entrained molecular oscillation of *per1* in Chica exists in fish from the natural cave (Fig. 6.27). It can be concluded that the lack of rhythmicity is not due to the absence of sufficiently strong entraining signals within the cave, because they are provided by the circadian activity of an enormous number of bats roosting there. Rather, the core clock mechanism is hypothesized to be tonically repressed or damped as a consequence of overactivation of the light input pathway (Beale et al. 2013). Possibly the differing *per1* expression levels in the wild and under laboratory conditions are indicative of an existing variability resulting from the phylogenetically young age of the VEP Chica cave fish. Enhanced variability is a characteristic of regressive traits and also exhibited by other regressive features like the eye, body pigmentation, or fright reaction in the Chica fish (see Sects. 6.17, 6.21.3, 7.8.1).

In previous studies, behavioural activity of cave animals like amphipods and salamanders has been examined to analyze whether circadian rhythms are lost. It is therefore an interesting possibility that in cave fish the molecular circadian rhythm is somehow uncoupled from the control of locomotor activity (Fig. 6.28) (Beale et al. 2013). For example, as an output of the circadian clock, *Astyanax* surface fish show opposite circadian rhythms with respect to their activity in different water depths (Erckens and Martin 1982a, b). In nature, this behaviour is probably performed as a means to avoid predators, being different during the day and night. Activity close to the water surface peaks in the D-phase, whereas activity near the bottom is in the L-phase. Based on the study of this behaviour, two endogenous time-measuring mechanisms were found to be responsible in the surface fish (Erckens and Weber 1976; Erckens and Martin 1982a, b). Firstly, a foreign-sustained (passive) one can be induced by LD (1:0 lux) of any period length. It damps out in DD within one or a few cycles with decreasing amplitude, but unchanged frequency. Secondly, free-running circadian activity rhythms, which are obviously under the control of a self-sustained clock system, exhibit in DD, LL (constant light), and also in LDs that deviate from 24 h LDs by several hours. The

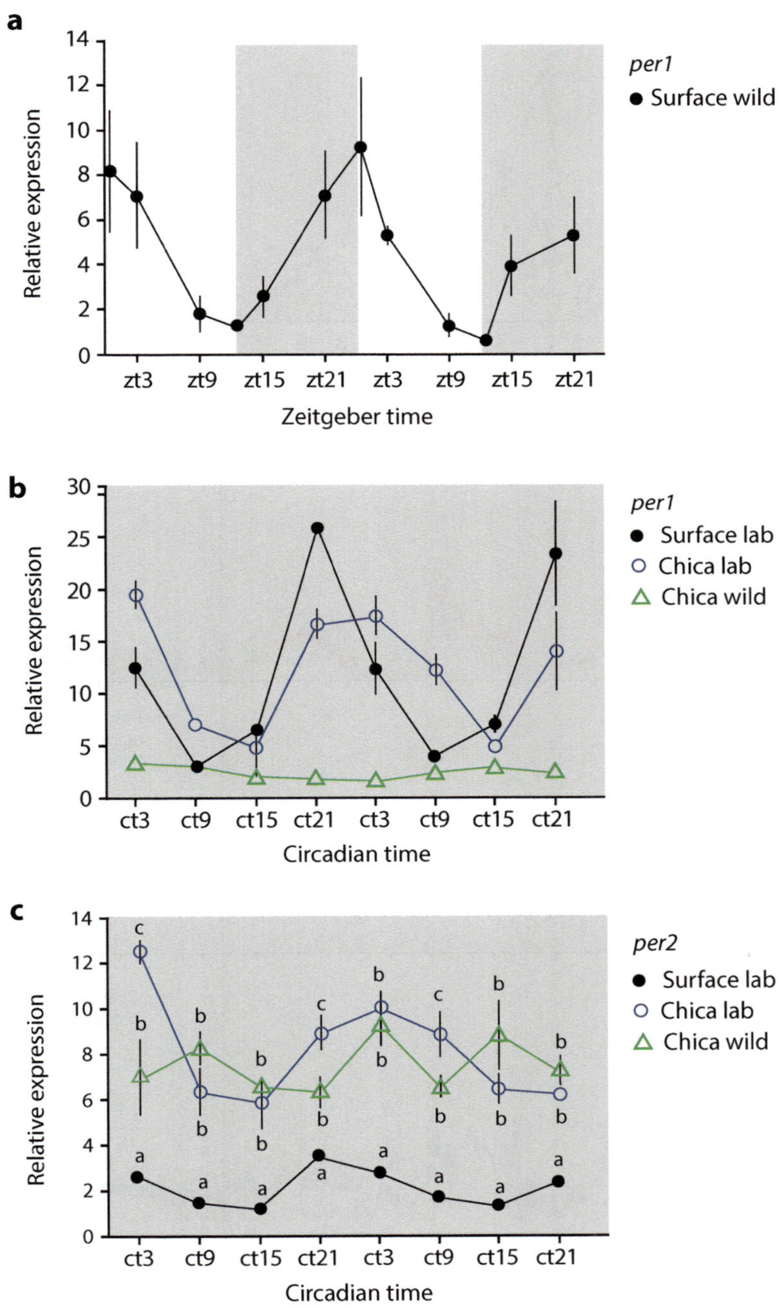


**Fig. 6.27** Circadian clock gene (*period 1* and *2*) expression in fin samples from laboratory-kept and wild *Astyanax* surface fish and Chica cave fish. *White* and *grey* areas indicate light and dark periods. (**a**) Surface fish from the wild, (**b** and **c**) surface fish and Chica fish entrained in the lab and Chica fish from the wild (Different *lower case letters* in (**c**) indicate significant differences [$p < 0.05$] between comparisons) (adapted from Beale et al. 2013)

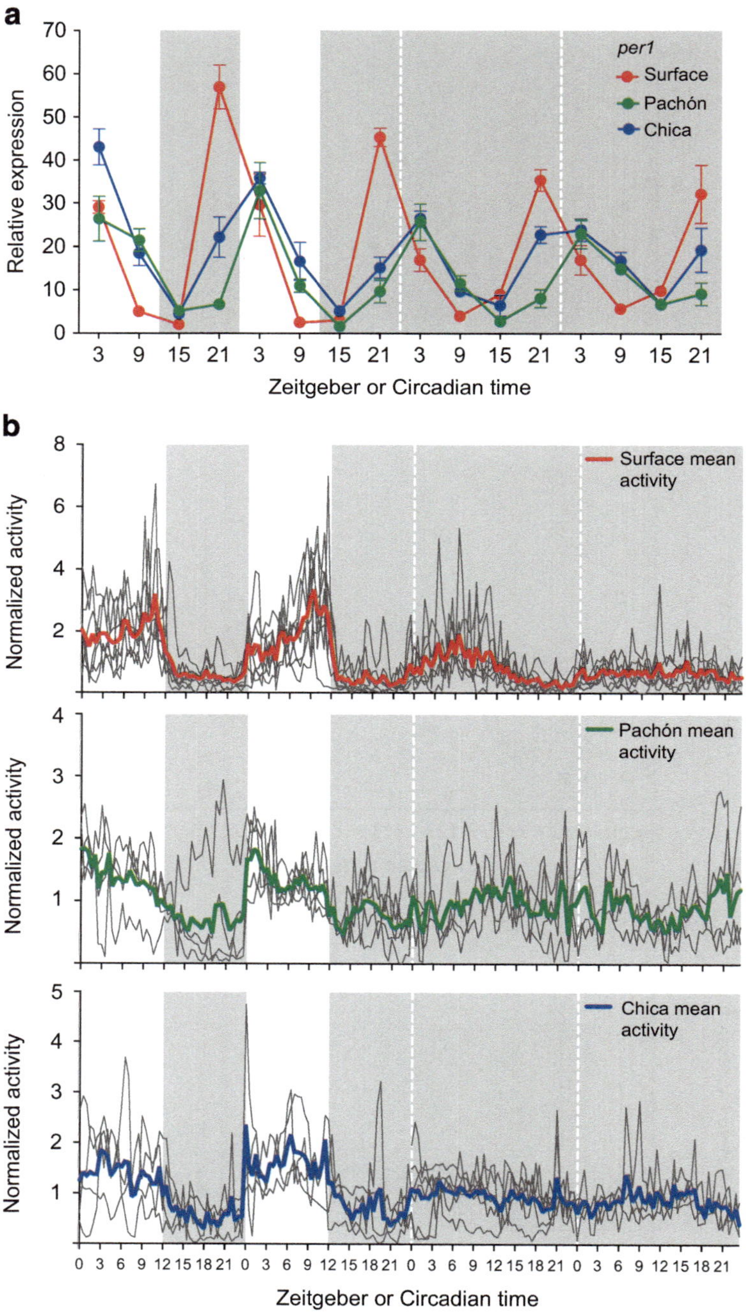


**Fig. 6.28** Molecular and locomotor analyses of the *Astyanax* surface fish, the SEP Pachón and the VEP Chica cave fish in LD and DD. (**a**) Expression of clock gene *per 1*, (**b**) swimming activity of

free-running activity was mostly expressed only in the surface activity, while the bottom activity was arrhythmic. Compared with the surface fish, the Pachón cave fish has similar, but much less pronounced circadian rhythms in surface- and bottom-directed activities in LD. Its time-measuring ability was found to be regressive but the circadian clock system has not been reduced (Erckens and Martin 1982a, b; Erckens and Weber 1976).

The study of swimming behaviour corroborates these findings (Beale et al. 2013). Surface fish show a clear, diurnal circadian rhythm in LD (23.95 ± 1.01 h), which persists on the first day in DD (29.77 ± 2.49 h), but then becomes arrhythmic on the second day in DD (Fig. 6.28). Pachón and Chica cave fish also show significant circadian rhythms in LD (23.95 ± 0.82 h and 25.6 ± 1.30 h, respectively), but after entry into DD, both populations, like the surface fish, are behaviourally arrhythmic and show substantial variation between individuals. Cave fish activity remains at median-to-peak levels in DD, whereas surface fish activity drops closer to basal lower levels (Beale et al. 2013).

For the SEP Tinaja cave fish, it was shown that it retains its ability to show light entrainable circadian rhythms of locomotor activity (Caballero-Hernandez et al. 2015). It was found to persist during the 10 days before being exposed to light cycles and some of the specimens were able to exhibit phase control with the previous LD when released in DD, indicating a true entrainment. Like for sleep loss (Duboué et al. 2012) (see Sect. 6.10), Caballero-Hernandez et al. (2015) propose for the Tinaja cave fish an endogenous behavioural adaptation to an environment with extremely low food resources, in order to stay in a quasi-permanent state of alert at all times, but with potential induction of cycles when environmental conditions change. The entrainment responses vary between different Tinaja individuals indicating that the possible fading of the rhythmic circadian activity is variable and may occur differentially among specimens of the same population.

Sleep and activity periods also provide examples for the uncoupling from circadian rhythmicity (see Sect. 6.10) (Duboué et al. 2012). The same was reported to happen to the circadian rhythm of the aerobic metabolism, which is uncorrelated to locomotor activity in the Pachón cave fish (see Sect. 6.9) (Moran et al. 2014).

The conservation of circadian rhythmicity during regressive evolution was also found in the eye rudiments of embryonic *Astyanax* cave fish. Light-entrained shifting rhythms of pigment granules take place in the retinal pigment epithelium of the degenerating eyes of the SEP Pachón and the VEP Chica cave fish (Espinasa and Jeffery 2006). After LD, the granules continue shifting under DD for two cycles. Entrainment is probably by light, perceived by photosensitive ganglion cells of the inner nuclear layer, which, in contrast to the visual cells (including

**Fig. 6.28** (continued) the surface fish, the Pachón cave, and the Chica cave fish. *White* and *grey* shaded areas indicate light and dark periods, *DD* continual darkness, *LD* alternating periods of light and dark, *SEP* strongly eye- and pigment-reduced, *VEP* variably eye- and pigment-reduced (adapted from Beale et al. 2013)

outer nuclear and outer plexifom layers), is retained in part of the eyes of the Pachón cave fish (see Sect. 6.21.1) (Peirson et al. 2009; Wilkens 1988, 2007).

Together, these results show that *Astyanax* cave fish possess a functional circadian clock, but one that is less robust than that of the surface fish. In contrast with the reduced eye, the circadian clock seems largely conserved in *Astyanax* cave fish. This is confirmed by transcriptome studies of surface and cave *Astyanax* (Gross and Wilkens 2013; Hinaux et al. 2013). In the Somalian cave fish *Phreatichthys andruzzi*, more severe regressive disturbances were revealed. Mutations in two circadian photoreceptors sequences, melanopsin and Teleost Multiple Tissue-opsin (TMT), have led to the loss of light entrainment of the clock ("blind clock", Cavallari et al. 2011). An examination of the TMT-opsin sequences between *Astyanax* surface and Pachón cave fish did not show such alterations (Beale et al. 2013). In *P. andruzzi* it was found, however, that feeding time has replaced light as a zeitgeber and a food-entrainable pathway exists. An infradian 47-hour free-running period is exhibited, by which it could be insinuated that the 24-hour period in principle still exists but in a doubled period to conserve homeostasis.

In contrast to *P. andruzzi*, visual light can still be used to entrain the circadian clock in a 24-hour rhythm via the pineal in *Astyanax* cave fish. As in most cave species, it can only be speculated about the nature of the zeitgeber by which the circadian clock is entrained in *Astyanax* cave fish in the natural continuously dark environment. It was proposed to be food-entrained, like in *P. andruzzi*, because bats roost in the caves (Mitchell et al. 1977; Wilkens and Burns 1972). However, although the Chica cave fish live in a cave that is settled by thousands of bats, no entrained molecular oscillation of *per1* was detected in the wild. This observation is explained by the hypothesis that the oscillator possibly does not run under cave conditions (Beale et al. 2013). It is suggested that in order to regulate homeostasis, the *Astyanax* cave fish cannot afford to lose the basic principles of circadian rhythmicity such as the rhythmic production of melatonin in the pineal, which consequently is submitted to stabilizing selection. In contrast, traits like surface water- and bottom-directed activity, which have lost the biological function, are no longer subjected to selection and therefore this behaviour as well as its rhythmicity may become reduced. Induced rhythms deviating from 24 hours were shown to disappear in DD immediately without any post-oscillations in the Pachón cave fish (Erckens and Martin 1982a, b). Furthermore, the circadian rhythmicity of aerobic metabolism found in surface *Astyanax* is no longer exhibited in the Pachón cave fish (Moran et al. 2014, 2015) and sleep rhythmicity has changed, too (Duboué et al. 2011; Erckens and Martin 1982a, b). These traits have become adaptive in *Astyanax* cave life, however.

Several light-induced genes, including *cry1a* (cryptochrome) and *per2*, are known to be critical for light resetting of the circadian pacemaker (Albrecht and Ripperger 2009). It was therefore proposed that the basally increased levels, particularly of *per2* in cave fish, are likely to lead to the extremely reduced *per1* expression detected under natural conditions and stop the clock (Beale et al. 2013). The authors furthermore argue that these molecular changes in the light input pathway contribute to the phase differences they observed between surface and cave populations in the laboratory, and especially to the reduced amplitude of the core clock oscillation. They conclude that the light input pathway exists in a more activated state in cave populations, as if the fish were actually experiencing constant light rather than the perpetual darkness of the cave, and it is hypothesized that this might be responsible for a selective advantage. The data analyzed in wild Chica specimens show complete repression of clock function (Fig. 6.27), while in the Pachón and the Chica cave fish, two genes involved in DNA repair, *CPD phr* (cyclobutane pyrimidine dimer-photolyase) and *ddb2* (damaged DNA-binding), show raised basal levels of expression. In field samples of wild Chica cave fish, the level of *CPD phr* and *ddb2* expression is increased over both the surface fish and laboratory cave strains, although they were never exposed to light. It is speculated that CPD photolyase in wild cave populations might have a functional role in DNA repair within the complete darkness of the cave and that altered expression of light-inducible genes provides a selective advantage to cave fish. This speculation relies on the finding that CPD photolyase can repress the mammalian circadian oscillator (Chaves et al. 2011). The high levels of CPD photolyase in the cave fish are assumed to have the same effect and are claimed to be responsible for the highly repressed levels of clock function in the Chica cave fish. By tonically activating light-dependent signalling pathways and increasing DNA repair activity, individuals in the cave would reduce deleterious mutational events, whereas the circadian oscillator would be damped as a byproduct. However, this hypothesis is predicated on the questionable assumption that the mutation rate would be higher in cave environments and above all would provide some kind of indirect selection pressure being responsible for the regression of the circadian clock.

## 6.17 Fright Reaction

Von Frisch was the first to describe that the European minnow (*P. phoxinus*) when injured releases flight in conspecifics. This alarm or fright reaction is found in Ostariophysan fish like *Astyanax* and is caused by an alarm pheromone, the *Schreckstoff*, produced in the skin epithelial alarm substance club cells (von Frisch 1941; Pfeiffer 1960). These cells do not have pores and the alarm substance is therefore not released voluntarily but after the epidermis has been hurt by an aggressor or a predator (Chivers et al. 2007; Fricke 1988; Peters et al. 1990).

To analyze this behaviour the experimental fish were subjected to an acclimatization time of 2 weeks, during which once every day 20 ml of tap water was squeezed with a glass pipette below the water surface and dry food flakes were immediately offered on the water surface. During the subsequent experimental period, water containing alarm substance prepared from an *Astyanax* surface fish was introduced in the same manner (Fricke 1988).

After the introduction of the alarm substance, the surface fish immediately exhibited a fright reaction, which consists of swimming a zigzag path, rapid swimming, hiding, and avoidance of the area where the alarm substance

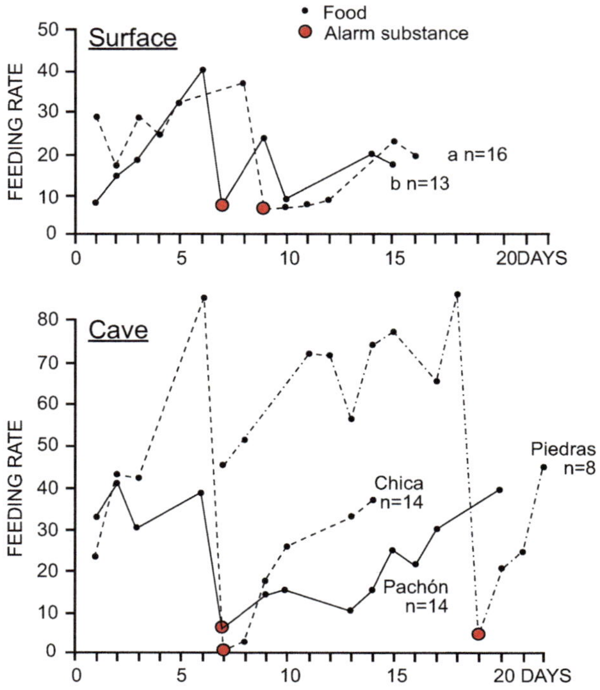


**Fig. 6.29** Feeding rate in two *Astyanax* surface fish [(**a**) Cenote Dzibilchaltún (Yucatán) and (**b**) Rio Teapao (Tabasco)], the VEP Chica and the SEP Pachón and Piedras cave fish before and after introducing a conspecific alarm substance at the water surface (*black dot* introduction of food, *red dot* introduction of conspecific alarm substance, feeding rate = snaps at food per 5 min). *SEP* strongly eye- and pigment-reduced, *VEP* variably eye- and pigment-reduced (adapted from Fricke 1988)

was set free. Furthermore, the feeding rate decreased significantly for several days (Fig. 6.29).

In the SEP Piedras and Pachón cave fish, only one of these reactions, namely avoidance, is exhibited whereas all others are lost. The specimens strictly do not swim into a layer of approximately 6 cm below the water surface, while they continue feeding on the ground. The VEP Chica cave fish shows the same behaviour but differs to a degree from the SEP Piedras and Pachón cave fish. A few of the Chica specimens still show reactions only exhibited by the surface fish. In all surface and cave populations, the fright reactions were found to persist over a period of nearly 6 days (Parzefall and Fricke 1991). As the alarm substance is not stable in water (Pfeiffer 1982) the persistence of fright reactions seems to be due to memory (Fricke 1988) (Fig. 6.29).

The alarm substance has not been altered in the cave fish and is perceived by olfaction (Chivers et al. 2007). However, in the surface fish, vision obviously plays an additional role, because specimens that have not noticed a danger by olfaction are alarmed by the behaviour of their conspecifics. The differences in behaviour between surface and cave fish can be explained by the absence of light in the caves. Whereas surface fish at daylight can be visually alarmed by conspecifics swimming rapidly or in a zigzag path, in cave fish visual alarm signals are useless and became completely reduced in the SEP cave fish. Only in the Chica cave fish a few of the

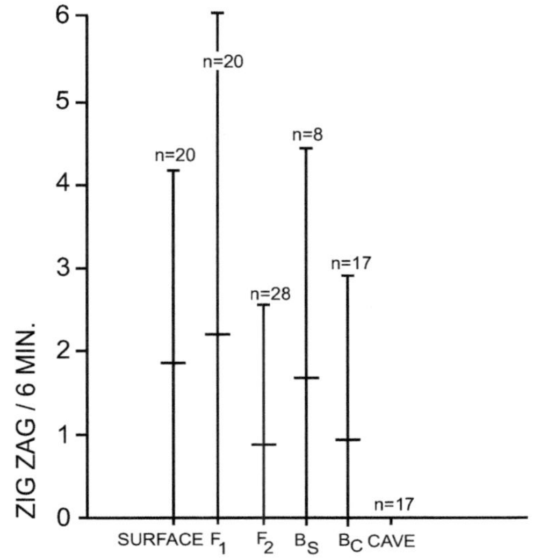


**Fig. 6.30** Means and ranges of the fright reaction zigzag movement in the *Astyanax* surface fish, the SEP Piedras cave fish and their F1, F2, and backcrossings demonstrate additive polygenic inheritance. $B_S$ backcrossing to surface, $B_C$ backcrossing to cave fish, *SEP* strongly eye- and pigment-reduced (adapted from Parzefall and Fricke 1991)

visually effective elements still occur. This is concordant with the variability of other regressive traits like the eye exhibited in this population.

It is proposed that, in contrast to visual releasers, only avoidance still has a biological function in cave fish. By this behavioural trait, protection is provided against foraging conspecifics or terrestrial predators like bull dog bats (Noctilionidae), which catch fish at or in a narrow layer below the water surface. Crossings between the surface and the VEP Piedras cave fish have revealed that the zigzag swimming behavioural trait relies on polygenic inheritance, because the mean of the F2 crossing is intermediate and the distribution of the intensity of this trait ranges between surface and cave fish. The means of the F1 and the backcrossing to the surface fish equate to that of the surface fish, indicating that epistatic gene effect is involved (Fig. 6.30) (see Sect. 6.23).

## 6.18 Schooling and Shoaling Behaviour

By definition, groups of fish aggregating with other fish of the same species are termed shoals, and such groupings are called schools when they synchronize and polarize swimming activities (Pitcher 1983). Under daylight conditions, surface *Astyanax* fish may alternate between these two types of social grouping but also may become territorial under conditions of limited space (Parzefall 1979, 1983, 1993; Gregson and Burt de Perera 2007). Schooling tendency is measured by the time a fish is following a moving model school. Shoaling is measured by distance and time one or more specimens aggregate to each other (Kowalko et al. 2013b; Parzefall 1992).

In darkness, surface fish are unable to form shoals or schools (Fig. 6.31a, b) (Parzefall 1993). Similar to surface fish in darkness, the visionless Pachón, Tinaja,

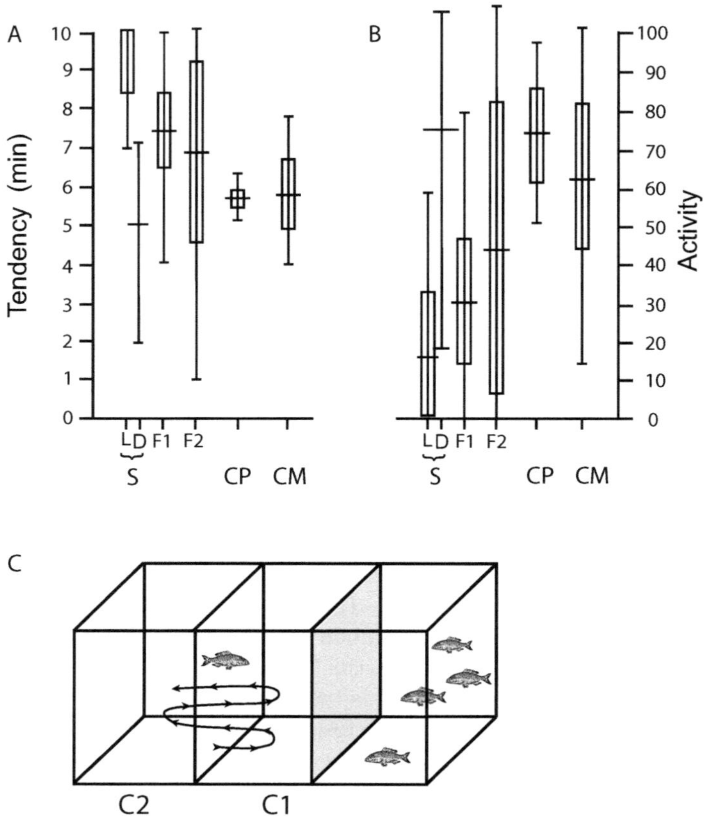


**Fig. 6.31** Tendency to shoal (**a**) and swimming activity (**b**) of the *Astyanax* surface fish (*S*) and the SEP Piedras cave fish (*CP*), the F1 and F2 crossings between them, and of the VEP Micos cave fish (*CM*). F2 hybrids and Micos specimens are selected for a good optomotor response and functional eyes, mean of $n = 10$ specimens in each group, *box* standard deviation, *vertical line* range, *L* study at light, *D* study in darkness. Experimental set up (**c**): tendency to aggregate to a group of fish = time staying in C1 (measured in minutes/10 min); swimming activity = number of changes per 10 min between chambers C1 and C2; *SEP* strongly eye- and pigment-reduced, *VEP* variably eye- and pigment-reduced (adapted from Parzefall 1993)

Piedras, and Molino cave fish do not aggregate to shoal or to school, but continuously swim without contact throughout the whole space available. This behavioural trait loss has evolved convergently in the different separate cave populations (Kowalko et al. 2013b; Parzefall 1993).

A study of specimens of the VEP Micos cave fish equipped with good vision revealed that the tendency to aggregate to a shoal is as low as that of the SEP cave fish (Fig. 6.31a, b). The variability of the tendency to shoal is about half that of the F2 crossing between surface and cave fish, but much higher than in the SEP Piedras cave fish (Parzefall 1993), which could be explained by the more recent origin of the VEP Micos compared with the SEP Piedras cave fish.

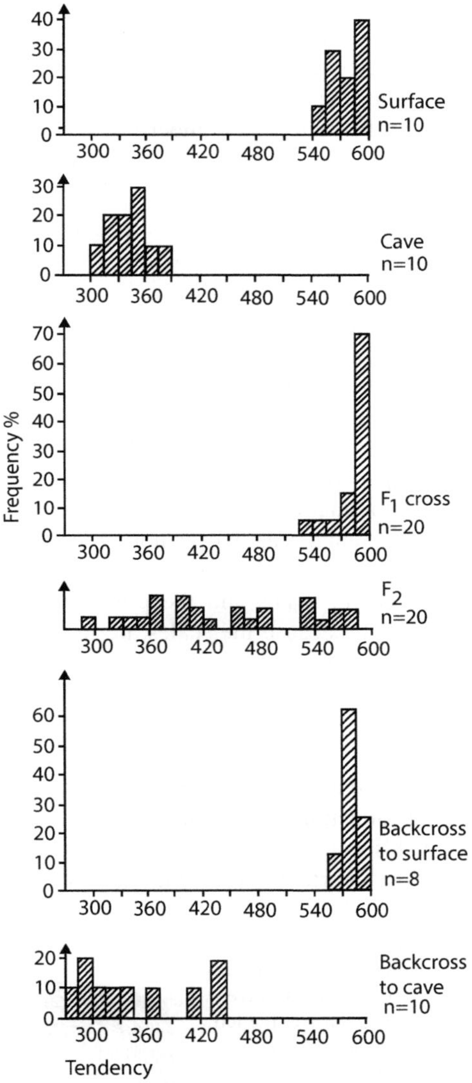


**Fig. 6.32** The distribution of the tendency to shoal (measured in seconds) of the *Astyanax* surface fish, the SEP Piedras cave fish and their F1, F2, and backcrossings indicate polygenic inheritance and epistatic gene effect. *SEP* strongly eye- and pigment-reduced (adapted from Parzefall and Fricke 1991)

Loss of vision in the cave fish is not solely sufficient to explain the reduction of shoaling and schooling behaviour. Also, the genetic basis of this behaviour is deteriorated. This was revealed by the distribution of the tendencies to school or to shoal in the crossings between surface and cave fish selected for good vision (Fig. 6.32) (Kowalko et al. 2013b; Parzefall and Fricke 1991). The F2 hybrids distribute between both parental forms and that of the backcross to the cave fish between the cave fish and the mean of the F2 crossing. Some F2 hybrids with good vision do not even aggregate. This indicates that a polygenic mode of inheritance is responsible. In addition, however, the distribution curves of the F1 crossing and the backcrossing to the surface fish equate to that of the surface fish due to epistatic gene effect (Fig. 6.32) (see Sect. 6.23) (Parzefall and Fricke 1991). QTL analysis has revealed that loss of shoaling behaviour in the *Astyanax* cave fish is based on at least two loci and that it has independently evolved from eye loss. Furthermore, neither the increased number nor size of sensory lateral-line free neuromasts are correlated with loss of shoaling behaviour (Kowalko et al. 2013b).

## 6.19 Scales

Fish scales provide protection against intra- and interspecific interaction and against parasites. Several cave fish species are characterized by their regression or even total loss (Banister 1984; Banister and Bunni 1980; Berti et al. 2001; Freyhof et al. 2016).

The body of *Astyanax* surface and cave fish is covered by the cycloid type of scales and their number along the lateral line is correlated to that of their vertebrae. In comparison with the surface fish, the scales of the cave fish show some characteristics of reduction. For example, in the Pachón cave fish they are shorter (Fig. 6.33) and the cranial scale part is not as deeply anchored in the scale pocket of the scale anterior to it, as is the case in the surface fish. Therefore, there is less overlap between the scales and the protective function is lowered. This effect is even intensified because the scales are less ossified and therefore are thinner. Furthermore, fixation of the scales of the cave fish is not as firm as in the surface fish, because they do not lodge as deeply in the scale pocket as they do in the surface form.

The partial reduction of the scales observed in *Astyanax* cave fish species is probably due to several facts. One reason may derive from the cave ecosystems being comparatively species-poor. For example, interspecific fights between different fish species competing for food do not occur because they mostly do not exist in the caves. Furthermore, the number of epidermal parasite species is lower, because

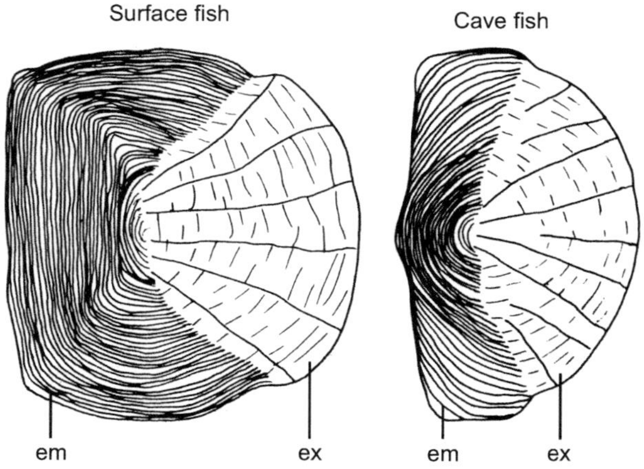


**Fig. 6.33** Comparative morphology of the scales of *Astyanax* surface and cave fish. Cranial parts to the left, *em* embedded part, *ex* exposed part

they cannot rely on alternate hosts in their life cycle. Also, intraspecific aggressive fights may occur less often.

## 6.20 Pigmentation

### 6.20.1 Melanophores

In *Astyanax* surface fish, males and females look similar and do not show sexually different or nuptial colouration. Their colour pattern mainly serves as camouflaging. An important part of body colouration is provided by the melanophore colour cells, which enable the fish to adapt to dark or bright background albedos (Fig. 6.34). This process may rapidly proceed within a few minutes by concentrating or expanding the melanin pigment granules within the melanophore colour cells (physiological colour change) or may take a longer time (several days) by additionally enhancing or lowering the number of melanophores (morphological colour change) (von Frisch 1911). During long-term background adaptation, alterations in melanophore morphology and density follow physiological colour change with the same factors involved in both phenomena (Sugimoto 2002). The most effective pigment-cell dispersing factor, the peptide α-Melanocyte-stimulating hormone (α-MSH), is thought to be the driving force responsible for the increase in melanophore dendricity and density by up-regulating melanogenic gene expression. Melanophore density in adult fish skin is likely to be maintained by a balance between differentiation and apoptosis (Fujii 2000; Sugimoto 2002). In surface *Astyanax* specimens being kept for a long time over a white background, the number of melanophores is statistically significantly lower compared with that developed over a black underground (Wilkens 1970b) (Fig. 6.55b). In darkness, the number of melanophores of the surface fish was revealed to be the lowest (Wilkens 1970b).

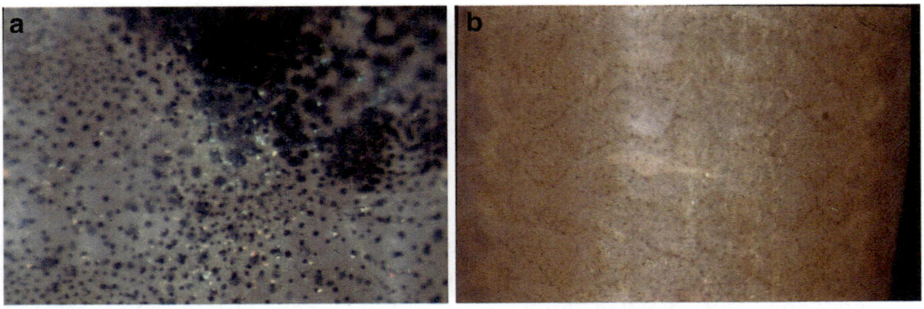


**Fig. 6.34** Sections of the back of *Astyanax* surface fish (**a**) exhibiting melanophores with dispersed and contracted melanin granules due to physiological colour change, and of the SEP Sabinos cave fish (**b**), in which they are reduced in number and contain less melanin due to the brown *Mc1r* mutation. The shiny spots are iridophores. *SEP* strongly eye- and pigment-reduced

Several cave species enhance the number of melanophores and melanin content when kept in daylight. For example, the hepapterid cave fish species genus *Rhamdia* achieve a brownish pigmentation, and the amphibian cave urodele *Proteus anguinus* may even exhibit intensively black body pigmentation. However, the SEP *Astyanax* cave fish does not darken when kept at daylight and exhibits the same number of melanophores as in total darkness. Contrary to this, the VEP Micos cave fish becomes darker at light because of the enhancement of melanophore number and melanin content.

The pattern of melanophore distribution on the body surface, with the highest density on the back and their number slowly declining down the sides, has not changed in the cave fish populations. The colour cells are still capable of performing a physiological colour change, too (Fig. 6.34). In the SEP cave fish, the pale colouration by an important part is achieved by the reduction of the number of melanophores. Among the SEP cave fish populations, the degree of reduction of the number of melanophores ranges between 70% and 85% compared with the surface fish (Wilkens 1970b, 1988; Wilkens and Strecker 2003). In contrast to the SEP cave fish, the VEP Micos cave fish develops the same number of melanophores as the surface fish when kept in daylight (Fig. 6.35) (Wilkens 1976). To assess the number of melanophores in the albinotic VEP Molino cave fish, crossing experiments with non-albinotic SEP cave fish were performed. As the albino gene is recessive, melanin production is then enabled and the melanophores become visible. As a result, it was revealed that the number of melanophores is about 50% of that in the surface fish (Fig. 6.36) (Wilkens and Strecker 2003). Thus, the number is higher than in the SEP, but lower than in the VEP Micos cave fish. The population-specific number of melanophores is already manifested at early embryogenesis by the number of melanoblasts, the precursors of the melanophores, which are developed in the neural crest of the cave fish (McCauley et al. 2004).

Crossings of the SEP Sabinos and Pachón cave fish show that the differing numbers of melanophores developed in different cave populations is mirrored by the means in the respective F2 and backcrossings to the cave fish (Wilkens 1970b; Wilkens and Strecker 2003). The method of inheritance of the melanophore number can be explained by polygeny and is characterized by epistatic gene effect (see Sect. 6.23, Fig. 6.55b). It relies on at least eight QTL or two genes found by classical crossing experiments (Protas et al. 2007; Wilkens 1988).

### 6.20.2 Brown Gene (*Mc1r*)

Besides the reduction of the number of melanophores, the paleness of the cave fish is additionally caused by partial reduction of the melanin content in the body melanophores. This so-called "brown gene" is found in the SEP Pachón, Yerbaniz, Japonés, Sabinos, Tinaja, Piedras, and Curva as well as in the VEP Chica cave populations. It is also responsible for lower melanin content in the retina pigmentary epithelial cells, due to which the eyes of preserved specimens no longer appear black but brown (Fig. 6.34b, Sadoglu and McKee 1969) (Table 4.1). In contrast, the

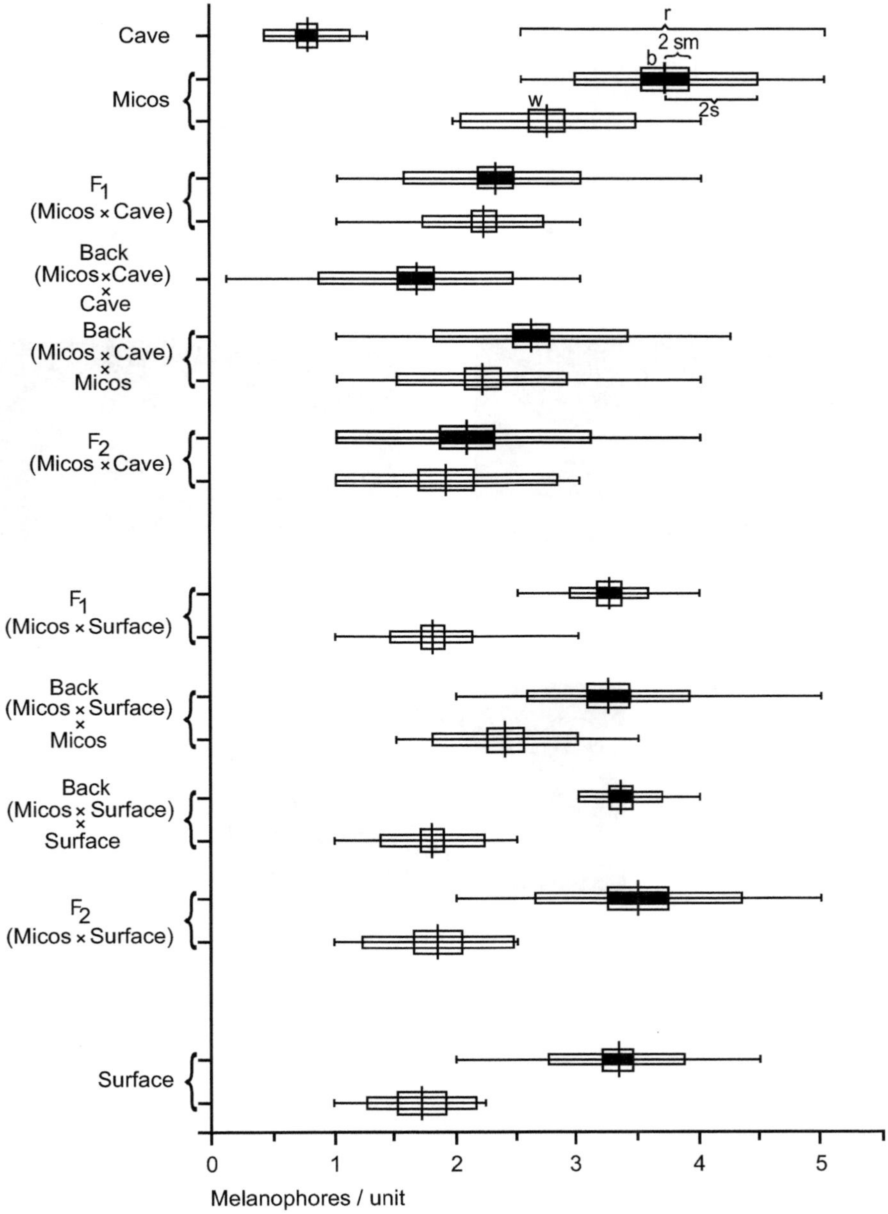


**Fig. 6.35** Mean, standard error (*sm*), standard deviation (*s*), and range (*r*) of the melanophoral densities of the VEP Micos cave fish over a white (*w*, *open*) or a black (*b*, *solid*) background. *Cave* SEP Pachón cave fish, *back* backcrossing, *Micos* Micos cave fish, *SEP* strongly eye- and pigment-reduced, *surface* surface fish, *VEP* variably eye- and pigment-reduced (adapted from Wilkens 1976)

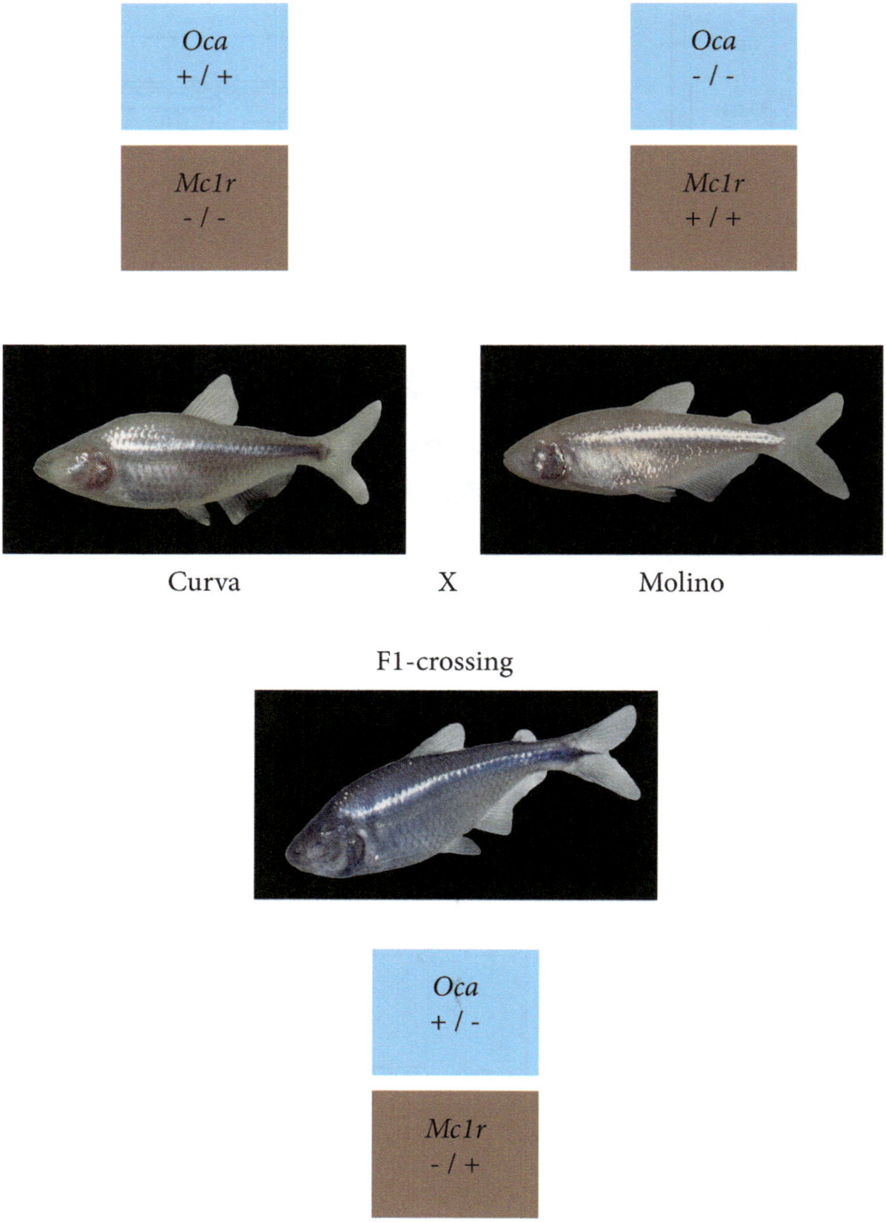


**Fig. 6.36** The F1-crossing hybrids of strongly reduced SEP cave fish like Curva with the albinotic VEP Molino cave fish show dark colouration, because the Molino cave fish still carries the unmutated wild allele of the brown gene (*Mc1r* gene), which solely does not manifest in this population because the Molino cave fish is albinotic and the recessive albino gene (*Oca2* gene), when homozygous, blocks melanin synthesis, *SEP* strongly eye- and pigment-reduced, *VEP* variably eye- and pigment-reduced

VEP Molino cave fish does not possess the "brown gene", but exhibits pale body colour, nonetheless, because it is albinotic and melanin production is completely blocked (Wilkens and Strecker 2003) (Fig. 6.36).

The "brown" mutation relies on a single recessive gene, which is inherited independently from those responsible for the number of melanophores as well as that for albinism (Sadoglu and McKee 1969; Wilkens 1970b). Crossing experiments revealed that the brown phenotypes manifesting in different SEP cave populations are all caused by mutations at the same locus and do not show genetic complementation in the crossings.

Using a candidate gene approach, the *Mc1r* gene (melanocortin type 1 receptor) was characterized as the likely locus controlling this trait. Sequence analyses revealed a 2-base-pair deletion in Pachón cave fish, which is predicted to cause a frame-shift leading to the introduction of a premature stop codon and to destroy normal functioning of the receptor protein. A mutation discovered in the Yerbaniz and the geographically closely linked Japonés populations causes a cysteine substitution, by which diminished function of the Mc1r protein in these populations is likely to be caused, explaining the presence of the brown mutation in these fish (Gross et al. 2009). Crossings demonstrate non-complementation of the brown phenotype in hybrid individuals derived from these two independent groups (Pachón vs Yerbaniz/Japonés), once again indicating that the same locus is responsible for reduced pigmentation in these fish (Gross et al. 2009). In the other SEP Sabinos, Tinaja, Piedras, and Curva cave populations, no differences in *Mc1r* coding sequence relative to the surface (wild type) populations were found, suggesting that these populations likely exhibit regulatory mutations leading to a decrease or loss of *Mc1r* activity (Gross et al. 2009; Stahl and Gross 2015). Thus, among 12 cave populations studied, the brown phenotype has arisen independently, mediated through different mutations of the same gene in the SEP Pachón, the Yerbaniz/Japonés, and in all other SEP cave populations that do not harbour a sequence alteration to the open reading frame. In that case, the brown phenotype may have evolved through the accumulation of sequence mutations affecting the 5′ regulatory region. Thus, the brown phenotype in *Astyanax* cave fish may have evolved through a combination of both coding and cis-regulatory alterations (Stahl and Gross 2015).

### 6.20.3 Albino Gene (*Oca2*)

Whereas the "brown gene" only causes a reduction of melanin, in those specimens carrying the albino mutation, the melanin of the melanophores is completely missing. The albino gene is recessive (Sadoglu 1957; Wilkens 1970b). It seems to be responsible for the inability to convert L-tyrosinase to L-dopa and melanin (Bilandžija et al. 2013; McCauley et al. 2004). Among the 27 *Astyanax* cave fish populations, albinism is rare and has only been found in the SEP Pachón, Yerbaniz/Japonés, and the VEP Molino populations. It appeared homozygous in Yerbaniz and Molino and for short time in Pachón (see Sect. 5.2.1). Melanophores in the albinotic populations are extant, but not visible due to the complete failure of melanin synthesis. However, they phenotypically manifest in the crossings with the surface fish (Sadoglu 1957; Wilkens 1970b; Wilkens and Strecker 2003). Albinism is linked to the *Oca2* gene (ocular and cutaneous albinism) and inherits independently from the genes responsible for the "brown" phenotype as well as from those for the number of melanophores (Wilkens 1988). On a molecular basis, in Pachón three differences from surface fish were found: two point mutations resulting in conserved amino acid substitutions each and a large deletion, which extends from within intron 23 through most of exon 24. Due to this deletion, the Pachón OCA2 protein would contain only part of intron 23 and lack most of exon 24. In Molino cave fish, a single change was found, namely a large deletion encompassing exon 21 that significantly shortens the OCA2 protein. Both differing coding sequence alterations affecting the Pachón and the Molino cave populations cause loss of function in the corresponding protein OCA2. From this can be concluded that the two cave populations evolved albinism independently through mutations at the same locus. It was also shown that the Yerbaniz/Japonés cave fish was deficient in *Oca2*. This seems, therefore, to be the third case of albinism detected, which separately arose, probably in a regulatory region (Protas et al. 2006; Protas and Patel 2008). A fourth case of albinism found in *Astyanax* cave fish appeared spontaneously in a captive stock of the VEP Micos cave population (Gross and Wilkens 2013), but has not, as yet, been detected in the Micos cave itself (Espinasa et al. 2014). Interestingly, it relies on the same loss of function alleles previously identified in the SEP Pachón cave fish (Gross and Wilkens 2013).

### 6.20.4 Guanine and Carotinoids

Besides the light-absorbing melanophores, light-reflecting, guanine-containing iridophores are developed as a second type of colour cell in the surface fish. They are developed in the lateral scales and protect the fish from being seen by predators attacking in the free water column from below. In addition, iridophores also exist in large numbers on the back, which helps camouflaging the surface fish over bright underground. This colour cell type is also submitted to regression in the cave fish. In the Molino cave fish, a slight reduction can additionally be found in the scales. However, the strongest reduction appeared in a captive stock of the Piedras cave fish, in which the guanine in the scales was totally lost. Because of this, the fish appear transparent although the scales are still present. This feature relies on a single recessive gene (Culver and Wilkens 2000) (Fig. 6.37a–c).

As a third type of colour cells, light-absorbing xanthophores are developed in the surface fish, which are particularly noticeable in the anal and caudal fins. The carotinoid pigments responsible for the colouration cannot be synthesized by the fish, but have to be ingested with food. A remarkable phenotype appeared in a captive stock of the SEP Piedras cave fish, the complete body of which was homogeneously coloured yellow, showing more intensive colouration in male than in female fish. It was suggested that this monogenic recessive mutant was

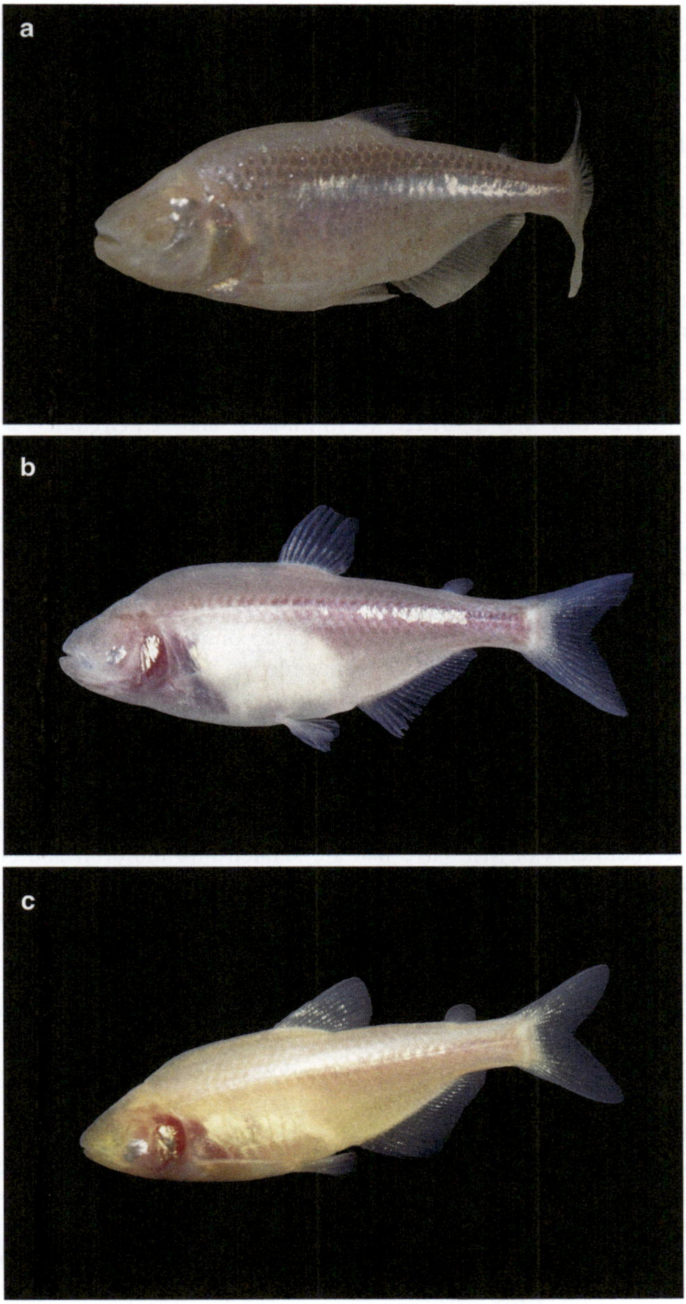


**Fig. 6.37** Two recessive monogenic colour mutations as yet not found in nature appeared in the SEP Piedras laboratory strains, most probably due to standing variation: besides the normal phenotype (**a**), specimens exhibiting total loss of guanine in the scales (**b**) and a yellow mutant (**c**) developed. *SEP* strongly eye- and pigment-reduced

unable to break down the ingested pigments and that they were stored in the fatty body tissue (Culver and Wilkens 2000) (Fig. 6.37a–c).

## 6.21 Eye

### 6.21.1 Morphology and Histology of the Eyes of the Surface and the SEP *Astyanax* Cave Fish

Eye reduction is a key characteristic of cave animals and has always fascinated scientists. In different cave fish species, all stages of reduction from the slightest decrease in size not yet afflicted by any structural defects to a degenerate rudiment deeply sunken into the head may be found (Fig. 6.38). Only a few species like the Somalian cave fish *P. andruzzi* (Cyprinidae) and *Uegitglanis zammaroni* (Clariidae) are known to have completely lost even the tiniest eye rudiment in adult specimens (Berti et al. 2001).

Whereas the eyes of the mostly troglophile surface sister species of cave fish are relatively small and adapted to perceive light in twilight or nocturnal environments, those of the diurnal surface *Astyanax* are large and well differentiated (Peters and Peters 1966; Wilkens 1988). They are characterized by a wide pupil and a voluminous lens, which can be pro- and retracted for accommodation by the lens muscle inserting on the distal tip of the falciform process. The retina is well developed, containing rods as well as single and double cones, the outer segments of which deeply extend between the cubic pigmentary epithelium cells. The chorioid or chorioidal gland is large, forming a horseshoe-shaped body of capillaries between retina and sclera. Guanine-containing light-reflecting layers build the argentea, improving light perception in shaded environments (Peters and Peters 1966).

The eye rudiments of SEP cave fish are sunken into the orbit and covered by epidermal body tissue, to which they are connected by a thin duct of tissue. Their diameter is about 10–20% of that of the surface fish. As concerns the various SEP cave populations, the eye is smallest in the Piedras and largest in Pachón, with Yerbaniz and Curva cave fish lying somewhere in between (Wilkens and Strecker 2003). The VEP Micos cave fish exhibit variable eye sizes, ranging from being almost as small as that of the F1 crossing between the SEP and the surface fish and as large as that of the surface fish (Wilkens and Strecker 2003). In the cave fish, eye size and degree of rudimentary structural differentiation are correlated (Wilkens 1970c). In adult SEP cave specimens, the largest and best developed eyes still possess a tiny rudimentary retina, whereas the visual sensory cells including outer nuclear and outer plexiform layers are lost (Fig. 6.39a). In contrast, the clearly separated ganglionic, inner plexiform and inner nuclear layers are extant and still connected to the brain by a thin optic nerve. Horizontal and Muller glia cells are developed. The retina rudiment attaches to the pigmentary epithelium, the cells of which are flattened and do not exhibit the typical cubic form exhibited in the surface fish. The extant vitreous body is almost completely separated by the ventral falciform process into a larger and a smaller chamber and the lens is always completely

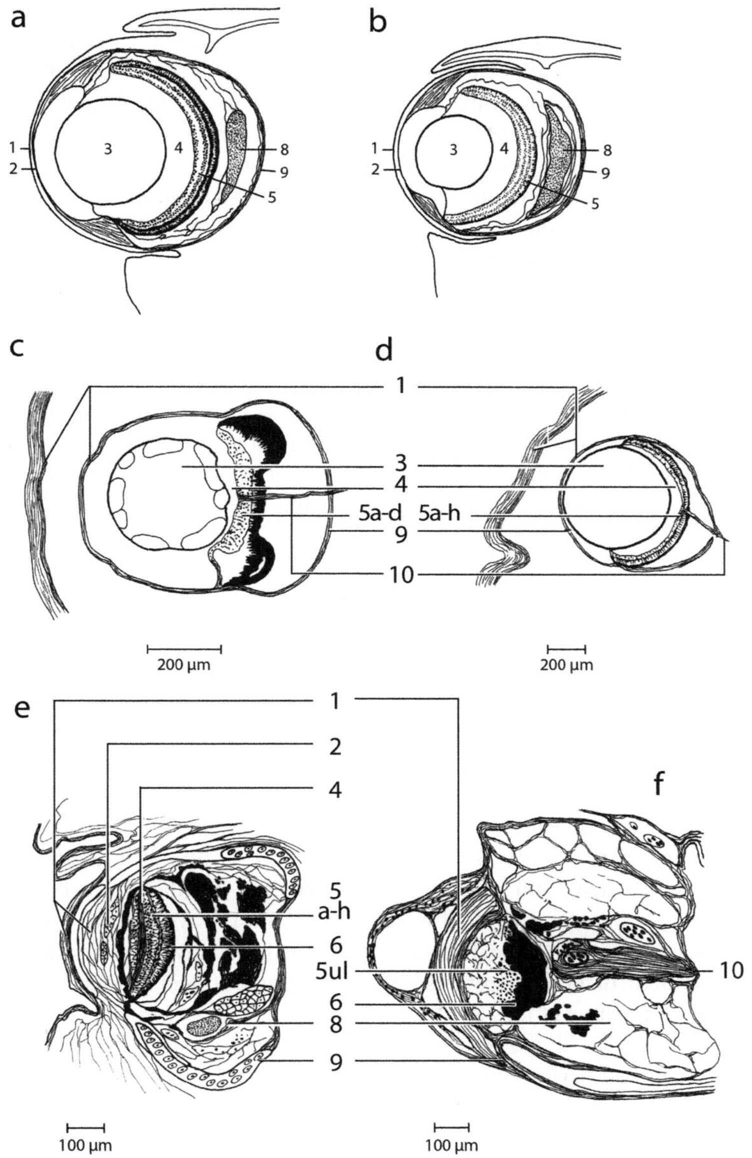


**Fig. 6.38** At its beginning, cavefish eye reduction is just a diminution of overall eyeball size and that of all single structures (**a** *Poecilia mexicana*, surface; **b** *P. mexicana*, cave). Only at later stages certain structures may get lost completely and considerable variability develops. In the cave cusk eel *Lucifuga spelaeotes* from the Bahamas, lens and retina are only reduced in size in the one from Grand Bahama (**d**), whereas in specimens from Abaco both show structural reduction (**c**). In the Oman cave fish *Garra barreimiae* the lens may be reduced whereas the retina still contains all characteristic layers of a functional eye (**e**) or the retina is no longer layered at all (**f**). Epithelial and scleral part of cornea (1), anterior eye chamber/spongiosum (2), lens/lens capsule (3), vitreous

missing in adults. The chorioid is reduced in size, too, and consists of connective tissue, blood vessels, and fragmented pieces of the argentea. A large part of the space enclosed by the sclera contains fat tissue. The rudimentary anterior eye chamber is entirely filled with the so-called spongiosum, which is spongious tissue deriving from the ligamentum annulare of the surface fish eye, which is located in the outer circular margin of the anterior eye chamber.

In the smallest eye rudiments of the SEP cave fish, specific characteristics are no longer identifiable. Neither layered retina, vitreous body, falciform process nor flattened pigmentary epithelial layers exist. Only remnant vessels of the choroidal gland are left in the voluminous space enclosed by the sclera (Wilkens 1970c). As well as the relatively well and the poorly differentiated eye rudiments, intermediate degrees of differentiation are developed (Fig. 6.39b) (Wilkens 1970c). Whereas the sclera of the surface fish is reinforced with an additional ring of bone derived from scleral ossicles (Yamamoto et al. 2003), scleral ossification has regressed convergently at least three times in the SEP Pachón, Sabinos, and Curva as well as in the VEP Chica and Molino cave populations (O'Quin et al. 2015). The eye coats are cartilaginous or consist of connective tissue (Fig. 6.39c).

### 6.21.2 Visual Pigments

Visual pigments function in light perception by transforming light energy into electrical (nerve) potentials. The study of the visual pigments of *Astyanax* has shown that the surface fish exhibits tetrachromacy and, in addition to a rod pigment, possesses one ultraviolet and violet sensitive, at least two rhodopsin, one rhodopsin-like, one blue, and two green cone opsin genes. The surface fish varies widely with respect to the ratio of Rhodopsin ($A_1$): Porphyropsin ($A_2$) chromophores, which divide into two groups with either 70% $A_1$ or 20% $A_1$ retinal in their cones. The basic spectral distribution of the cone pigments in the surface fish, with $\text{Longwave}_{\text{Green}}/\text{Longwave}_{\text{Red}}$ double cones and short wave sensitive (SWS) and possibly violet sensitive (VS) single cones is typical of many shallow water teleosts. The complement of cone pigments supports tetrachromatic colour vision (Parry et al. 2003; Yokoyama et al. 1995).

Sequence analysis of one red (*r007*) and two green sensitive opsin (*g101-2* and *g103-1*) genes was performed in the surface fish, the VEP Micos, and the SEP Pachón cave fish (Yokoyama et al. 1995). The coding regions exhibited 15, 7, and 16 polymorphic nucleotide sites, respectively. Eight out of these 38 changes are non-synonymous. Sixteen sites contained a specific type of nucleotide residue found only in one cave population. Intrapopulational nucleotide differences are

---

**Fig. 6.38** (continued) body (4), retina (5), retina rudiment (5a = inner limiting membrane, 5b = ganglionic layer, 5c = inner plexiform layer, 5d = inner nuclear layer, 5e = outer plexiform layer, 5f = outer nuclear layer, 5g = outer segments of visual cells, 5h = pigmentary epithelium, ul = unlayered), pigmentary epithelium (6), chorioid (8), sclera (9), optic nerve (10)

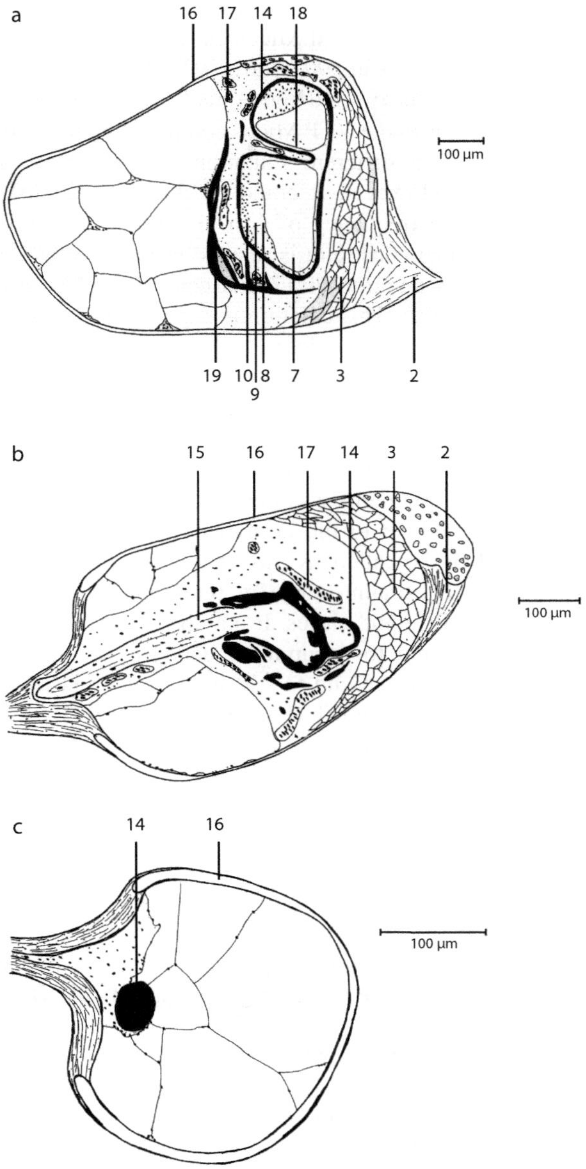


**Fig. 6.39** Schematic cross sections showing the range of structural variability of eye rudiments in adult SEP cave fish like Pachón and Sabinos: developed best (**a**), intermediate (**b**) and strongly reduced (**c**). (1) cornea (epithelial), (2) cornea (scleral) and/or connective duct, (3/3a) anterior eye chamber in part filled with spongiosum, (4) iris, (5) lens capsule, (5a) lens muscle, (6) lens, (7) vitrous body, (8) ganglionic layer, (9) inner plexiform layer, (10) inner nuclear layer, (11) outer plexiform layer, (12) outer nuclear layer, (13) visual cell outer segments, (14) pigmentary epithelium, (15) optic nerve, (16) sclera, (17) chorioid, (18) falciform process, (19) argentea; *R* retina unlayered, *SEP* strongly eye- and pigment-reduced (adapted from Wilkens 1970c)

the highest in the surface fish, decrease in the Micos, and are lowest in the Pachón cave fish. The lower levels in the cave fish are consistent with lower variability found in other molecular markers (see Sect. 4.3.3). However, in comparison, nucleotide diversities are higher between the surface and the Micos cave fish than between the surface and the Pachón cave fish in all three opsin genes. Of the

14 directed mutations in the cave fish, 12 belong to the VEP Micos and only 2 to the SEP Pachón. Pachón and Micos cave fish exhibit a high frequency of cytosine to thymidine changes in *r007* and *g103*, of which most have occurred in the Micos cave fish. The unexpected higher nucleotide divergence as well as the higher rate of C to T changes in the VEP Micos compared with the SEP Pachón cave fish could be explained by a different mutation rate for *r007* and *g103* in the two cave populations. However, it could be due to their different surface origin. In any case, the high frequency of C to T changes resembles that of pseudogenes and their functional constraint is relaxed in the cave fish. In addition to the nucleotide differences, a deletion of 12 consecutive nucleotides is exhibited in *g101* of all Pachón cave fish. This deletion is expected to induce strong deleterious effects on the ability to perceive colour.

Microspectrophotometry of *Astyanax* surface fish and its F2 crossing with the VEP Molino cave fish revealed that over half of the F2 hybrids possessed almost pure $A_1$ chromophores in their cones. The opsin genes remain as intact open reading frames in this cave fish (Parry et al. 2003). Nonetheless, a high incidence of "anomalous" pigments was detected. Such "anomalous" cone pigments are spectrally intermediate to the normal red and green ($L_R$ and $L_G$) pigments and are expressed by so-called hybrid genes. They are assumed to cause colour vision anomalies. The maintenance of the opsin genes found in the Molino cave fish might have different causes. It may be due to its phylogenetically young age with not enough time to accumulate deleterious mutations. However, as at least a red-like opsin was found to still briefly be transcribed in the embryonic retina of the SEP Piedras cave fish (Langecker et al. 1993), this may mean that opsin genes are under selection because they are required during ontogenetic retinal development and are responsible for the formation and folding of the outer segment disc membranes (Vinnikow 1982). Nonetheless, the majority of opsin genes are most probably on a course toward degeneration (Yokoyama et al. 1995).

### 6.21.3 Size and Histology of the Eye in the VEP Cave Fish and in the Crossings Between Surface and Cave Fish

#### 6.21.3.1 Crossings of the SEP Cave Fish

In the crossings between the *Astyanax* surface and SEP cave fish, different eye sizes ranging from small rudiments to normal sized eyes found in the surface fish develop (see Sect. 6.23, Fig. 6.55a). The F1 crossings are about intermediate with the means shifted towards the surface fish. The single eye structures like the retina or lens are intact and about half the size of that developed in the surface fish. Eye size variability is the same as in the surface fish. In relation to body size, the eye ball grows isometrically. Whereas the correlation between lens and pupil growth is isometrical, the relation of the lens to the eye ball is positively allometric. Because of this, juvenile hybrids have relatively smaller lenses and pupils than adult ones (Wilkens 1970a).

In the F2 and the backcrossing hybrids with the surface fish the growth of the individual lens and pupillary opening is positively allometric (Wilkens 1970a). Similarly, when considering hybrids of the same body size, the closely correlated lens and pupil sizes are positively allometric in relation to eye ball size. This shows that allometric growth correlations within the eye are determined by relative eye size—independently from whether they rely on the ontogenetic growth stage or a different individual genetic basis. However, particularly in the backcross hybrids with the SEP cave fish, the growth process of lens and pupil is submitted to the process of ontogenetic eye reduction. Thus, the mean relative lens and pupil sizes are relatively larger in juvenile backcross hybrids than in adults. In very small-eyed hybrids, the lenses may even become completely reduced and the closely correlated pupils are lost.

#### 6.21.3.2 VEP Cave Fish and Crossings

Whereas the SEP cave fish populations show strong reduction, the eyes of the VEP Micos, Chica, and Caballo Moro cave fish are less reduced, more variable in size, and, except for the smaller ones, externally visible. For example, eye size of the Micos cave fish approximately ranges between that of the F1 crossing of the surface and the SEP cave fish on the one hand and the surface fish on the other. When the distribution curve includes comparatively larger eyes, it exhibits bimodality (Fig. 6.55a). In the VEP Molino cave fish, the eye rudiment is sunken into the orbit and overgrown by tissue. Its size distribution partly overlaps with the larger ones of the SEP Pachón, but in contrast surpasses that of the SEP Piedras, Curva, and Yerbaniz cave fish (Wilkens and Strecker 2003).

The eye sizes developed in the crossings of the Micos cave fish with the SEP cave fish or with the surface fish depend on that of the respective parental Micos cave fish. The larger the eye, the higher the mean of the crossings. The variability of the lens size and the closely correlated pupil of the Micos cave fish itself are lower than in its F2 crossings with the surface fish or with the SEP cave fish. It is also lower than in the F2 crossings between the surface and the SEP cave fish (Fig. 6.40) (Wilkens 1976, 1988).

The mean eye size of the F1 crossing between the Molino and the surface fish converges with those of the F1 crossings between the surface and SEP cave fish. However, the variability is about three times higher, because eye balls smaller or larger than in the before-mentioned F1 crossings are developed. It can be concluded that this derives from comparatively high variability of eye genes.

#### 6.21.3.3 Histology of Eye Structures in the Crossings Between SEP Cave and Surface Fish and in VEP Cave Fish

Histological studies of the eyes in the various crossings between the surface and the SEP cave fish as well as in the VEP Micos and the Chica cave fish have revealed that, like in the SEP cave fish, size and degree of differentiation are in principal correlated (Wilkens 1972, 1988; Fack and Wilkens 1989). For example, the largest eyes are the best developed ones. They contain well differentiated retinas and transparent crystalline lenses and are comparable to those of the surface fish. In general, it was found that, on the one hand, the anterior eye chamber and the

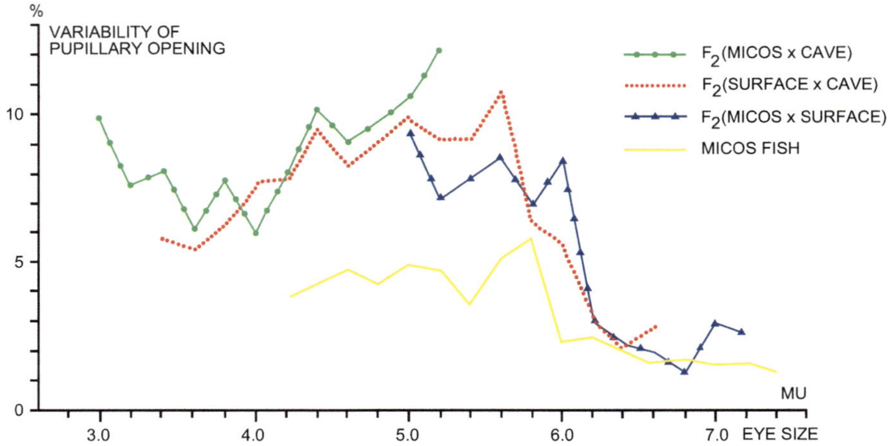


**Fig. 6.40** Variability of the pupil diameter of eye balls of the same size in the *Astyanax* VEP Micos cave fish (*yellow line*) and its F2 crossings with the SEP cave fish (*green line*), the surface fish (*blue line*), and in the F2 crossing between surface and SEP cave fish (*red dotted line*) at 2.5 cm standard body length, *MU* units of measurement, *SEP* strongly eye- and pigment-reduced, *VEP* variably eye- and pigment-reduced (adapted from Wilkens 1976)

ligamentum annulare, the thickness of the cornea, and the diameter of the pupil are correlated with the size of the lens. On the other hand, correlations exist between size and degree of differentiation of retina, pigmentary epithelium, lens muscle, and optic nerve. Both units seem to influence the volume of the vitreous body and the development of the sclera.

However, specific disturbances and developmental aberrations develop in the eyes of a lot of hybrids as well as in the VEP cave fish. In the extreme, eyes with large crystalline lenses may be combined with small retina rudiments (Fig. 6.41a) or well developed retinas may be found in eyes just containing rudimentary lens capsules (Fig. 6.41b). Furthermore, due to an oversized lens inducing the development of a vitreous body being too large for a retina being too small, part of the retina may be pressed as a funnel-shaped formation through the falciform process (Fig. 6.41c). Or, when the lens is too small to produce a sufficiently voluminous vitreous body adequate for a large retina, the retina may become undulated due to a lack of space (Fig. 6.41b). The same phenomenon was observed in chicken eyes, when the vitreous body was artificially prevented from accumulating (Coulombre and Coulombre 1963). Furthermore, in accordance with being part of the retina unit, a large lens muscle may exist while the functionally actually associated lens, which is the central part of the second subunit, only consists of a lens capsule (Fig. 6.41d). From this it can be concluded that two separate developmental sub-units exist within the eye, which show closer correlation within each and inherit independently.

Whereas the eyes of the hybrids between surface and SEP cave fish exhibit intermediate sizes between the parental forms, the crossings of the different SEP

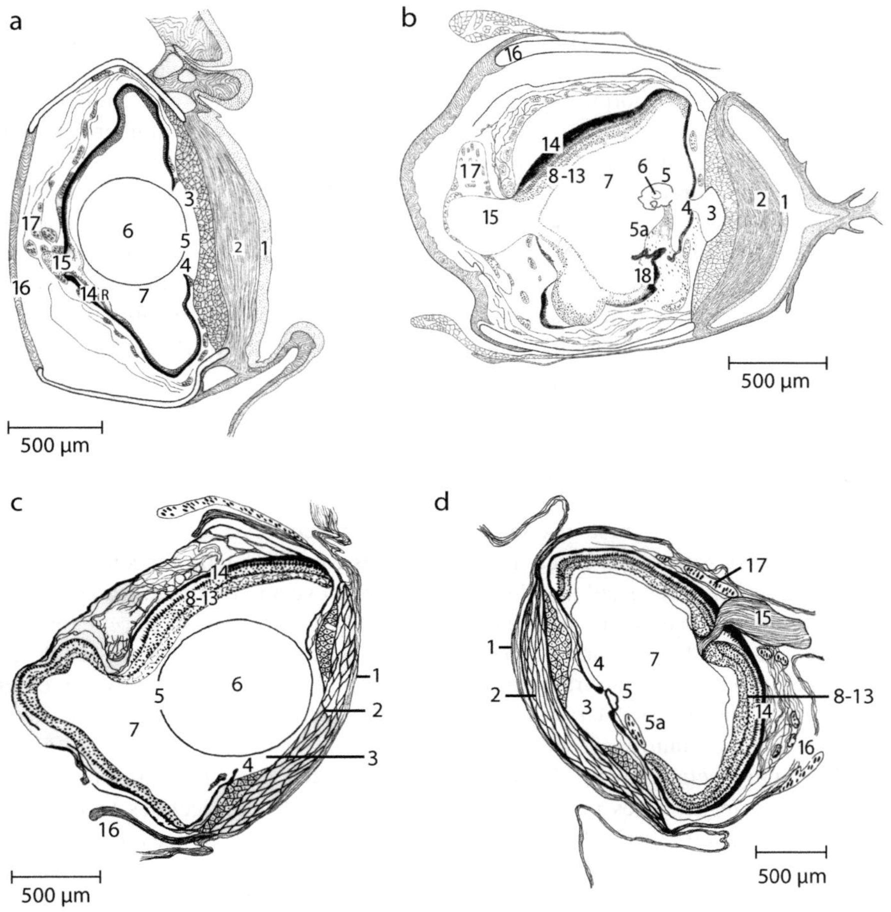


**Fig. 6.41** Schematic transverse sections of the eyes of two specimens of the *Astyanax* VEP Chica cave fish (a = large lens combined with undifferentiated retina, b = undifferentiated lens combined with lens muscle and partly well developed or distorted retinal parts) and two F2-hybrids between surface and the SEP Pachón cave fish (c = large lens combined with retina protruded into the chorioid and d = rudimentary lens combined with large lens muscle and well developed retina). *SEP* strongly eye- and pigment-reduced, *VEP* variably eye- and pigment-reduced (adapted from Wilkens 1972 and Fack and Wilkens 1989, for abbreviations see Fig. 6.39)

cave fish populations with each other as well as with the VEP Molino cave fish deviate from this. They may develop larger eyes than the parental cave fish. In the F1 crossings, specimens containing lens capsules may appear (Wilkens 1971, 1988, 2007; Wilkens and Strecker 2003). An exceptional situation is found in the F2 crossing between the Pachón and the Molino cave fish, in which "back to surface eyes" develop. Together with specimens exhibiting smaller eyes that are sunken into the orbital cavity, as in the parental generation, several of the adult F2 specimens develop externally visible eyes with large lenses and wide pupils (Fig. 6.42). Histological sections of the eyes from these latter specimens revealed the presence of large transparent crystalline lenses as well as retinas containing all optical layers with morphologically intact visual, horizontal, and Müller cells including pigment epithelium, optic nerve, and lens muscle. The number of visual cells is lower, though, and therefore the retina is not as thick as in the surface fish (Wilkens and Strecker 2003). This particular observation was only made in crossings with the SEP Pachón cave fish and can be explained by the Pachón cave fish exhibiting an eye genetic background less affected by regressive mutations than the other SEP cave fish (Wilkens and Strecker 2003). This could be caused by genetic introgression of an unknown VEP cave population, which is claimed to be responsible for the mitochondrial capture in the SEP Pachón cave fish (see Sects. 5.2.1, 5.4).

### 6.21.4 Ontogeny of Eye Development in the SEP Cave Fish

Cahn (1958) was the first to compare the ontogenetic eye development of two *Astyanax* cave forms, the Chica and the Sabinos cave fish, with the surface sister fish. These studies revealed that the larval cave fish eye and its single structures are not only smaller than in the surface fish, but already exhibit regressive development (Wilkens 2007; Wilkens and Meyer 1992). For example, neither of the lens cells will ever differentiate into transparent fibres nor will the visual cells develop outer segments. Thus, the SEP *Astyanax* cave fish do not start with a "complete eye", as is often inferred (Alunni et al. 2007; Jeffery 2001).

The ontogenetic growth of the eye ball of surface *Astyanax* starts positively allometrical in juveniles until about 4 cm body length and declines to negative allometry in adult stages after 6 cm body length is reached. In between, the eyes grow isometrically (Fig. 6.43) (Table 6.2). In the cave fish, four ontogenetic stages with specific histological and allometric growth characteristics can be differentiated (Figs. 6.43 and 6.44). At stage 1 (<4 mm body length), smaller primordia than in the surface fish are formed, but the relative growth correlations between eye and body size are similar (Table 6.2). In spite of its smaller size, the retina of the cave fish eye differentiates in the same manner as that of the surface fish. Usually the chorioid fissure closes and forms the falciform process. Lens fibre development, which in the surface fish begins at these early stages, fails in all strongly eye-reduced SEP and the VEP Molino cave fish, and the lens stays undifferentiated too (Wilkens 2007).

Occasionally the falciform process does not completely grow together in its ventral part. By this defect the vitreous body in cave fish is not able to properly accumulate, which is a prerequisite of eye growth (Coulombre and Coulombre 1963). Such cave fish eye rudiments cannot differentiate properly and remain smaller. In humans, this pathological defect is called coloboma iridis. In *Astyanax* cave fish it is an outlier (Figs. 6.12, 6.45, 6.46) (Wilkens 2007), but was falsely diagnosed as the ventral sector of the retina being reduced or missing and

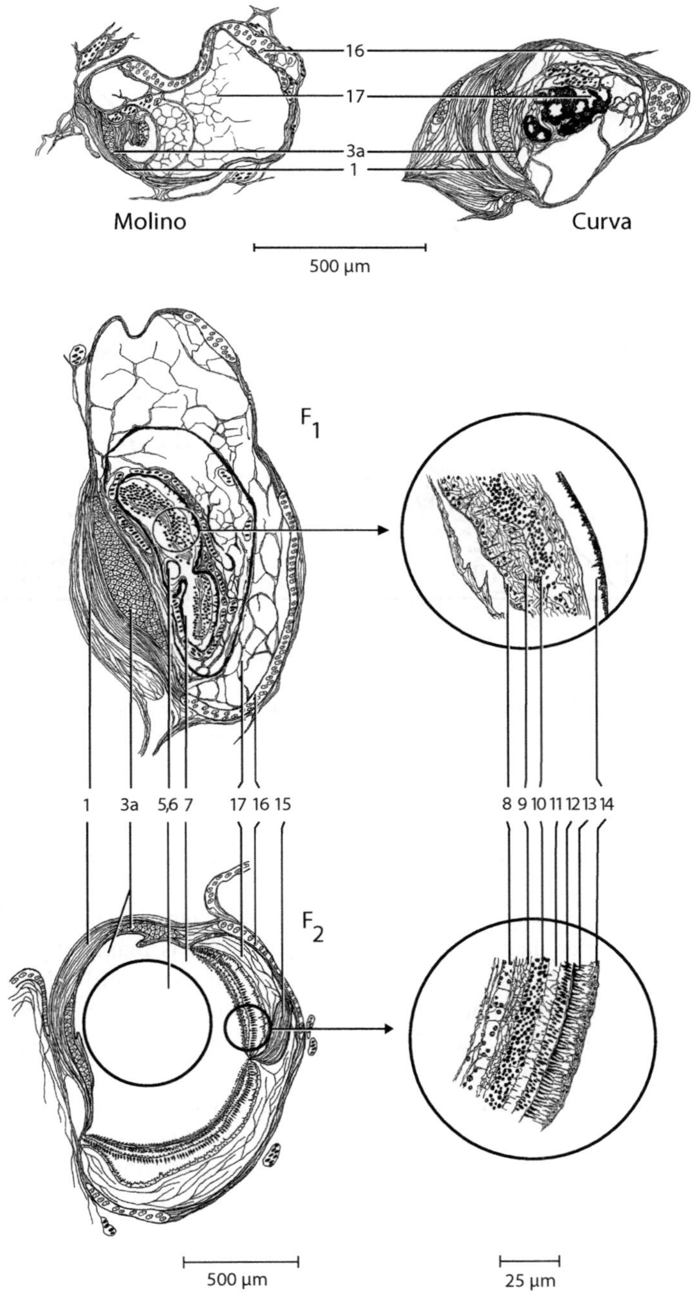


**Fig. 6.42** Transverse sections of the eyes of the VEP Molino, the SEP Curva cave fish and of F1-and F2-crossing hybrids between VEP Molino and SEP Pachón cave fish (for abbreviations see Fig. 6.39, adapted from Wilkens and Strecker 2003)

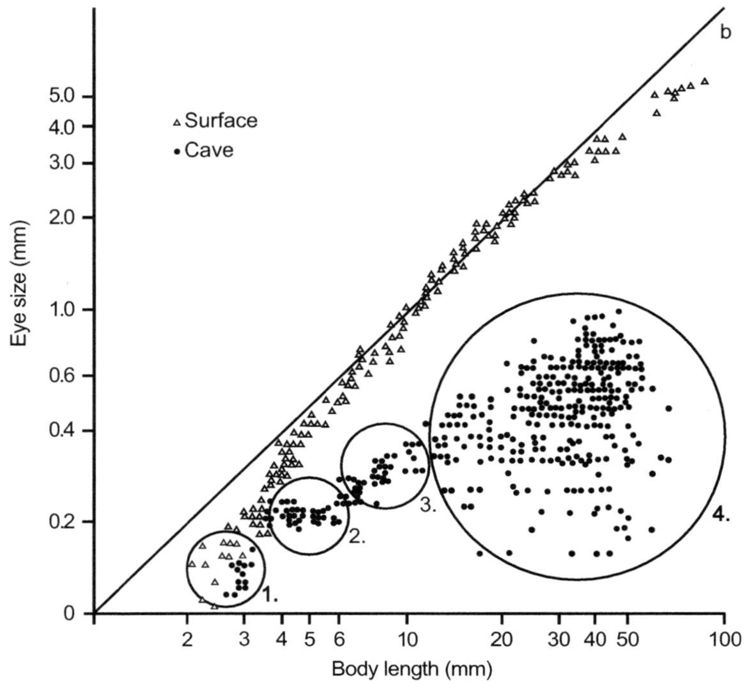


**Fig. 6.43** Allometry of ontogenetic eyeball growth in correlation with body size (double logarithmic) in *Astyanax* surface (*triangles*) and the SEP Sabinos cave fish (*dots*). *Circles* 1–4 indicate different growth stages in cave fish, *SEP* strongly eye- and pigment-reduced. (For allometrical coefficients see Table 6.2, adapted from Wilkens 2007)

interpreted as a general characteristic of eye regression (Jeffery et al. 2003; Pottin et al. 2011).

At the beginning of stage 2 (≥4 mm to <6 mm body length), the eyes of the SEP cave fish are smaller than the surface fish eye, which is maximally 0.4 mm in diameter. In the cave fish, average eye sizes are the largest in VEP Molino and the SEP Pachón cave fish (maximal sizes 0.32 and 0.28 mm, respectively), whereas the SEP Piedras, Sabinos, Curva, and Yerbaniz populations show smaller sizes (maximal size 0.25 mm). Most importantly, visual cell ellipsoids are rarely found in the retina and lamellar visual cell outer segments are always absent in the strongly eye-reduced SEP cave populations (Figs. 6.45a and 6.47a). As an exception, the VEP Molino cave fish, which possess relatively strongly reduced eyes compared with the VEP Micos and Caballo Moro cave fish, develop perfectly structured layers of visual cell ellipsoids containing rich accumulations of mitochondria in large numbers and even small primordia of lamellar outer segments (Fig. 6.45b).

A temporary stop or even a decline of eye growth and relative eye size can be observed during stage 2 in the cave fish (Figs. 6.43 and 6.44), whereas the surface fish eye continues growing. Such a temporary growth stop is also not found in the

**Table 6.2** Allometrical coefficient (*b*), coefficient of correlation ($r^2$) and number of individuals (*N*) of *Astyanax* surface and cave populations as well as F1, F2 and backcrossings (*R*) with Pachón at the different ontogenetic stages (adapted from Wilkens 2007)

| | Stage 1 (<4.0 mm) | | | Stage 2 (≥4.0 < 6.0 mm) | | | Stage 3 (≥6.0 < 15.0 mm) | | | Stage 4 (≥15.0 mm) | | |
|---|---|---|---|---|---|---|---|---|---|---|---|---|
| | *b* | $r^2$ | *N* | *b* | $r^2$ | *N* | *b* | $r^2$ | *N* | *b* | $r^2$ | *N* |
| Surface | 0.998 | 0.676 | 45 | 0.814 | 0.748 | 56 | 1.223 | 0.962 | 84 | 0.737 | 0.968 | 166 |
| $F_1$ | 1.026 | 0.344 | 22 | 0.972 | 0.848 | 22 | 1.321 | 0.987 | 16 | 0.768 | 0.856 | 31 |
| $F_2$ | | | | 0.647 | 0.380 | 203 | 1.000 | 0.787 | 135 | 0.717 | 0.607 | 338 |
| *R* | | | | 0.442 | 0.482 | 58 | 1.028 | 0.841 | 94 | 0.833 | 0.642 | 40 |
| Piedras | 0.923 | 0.703 | 30 | −0.112 | 0.029 | 34 | 0.830 | 0.840 | 38 | −0.480 | 0.302 | 34 |
| Curva | | | | 0.250 | 0.035 | 40 | 0.621 | 0.846 | 86 | 0.597 | 0.263 | 40 |
| Yerbaniz | | | | −0.760 | 0.154 | 35 | 0.657 | 0.715 | 60 | 0.663 | 0.545 | 25 |
| Pachón | | | | −0.140 | 0.034 | 24 | 1.102 | 0.931 | 22 | 0.271 | 0.240 | 34 |
| Molino | | | | 0.238 | 0.165 | 39 | 0.865 | 0.893 | 93 | 0.202 | 0.044 | 45 |

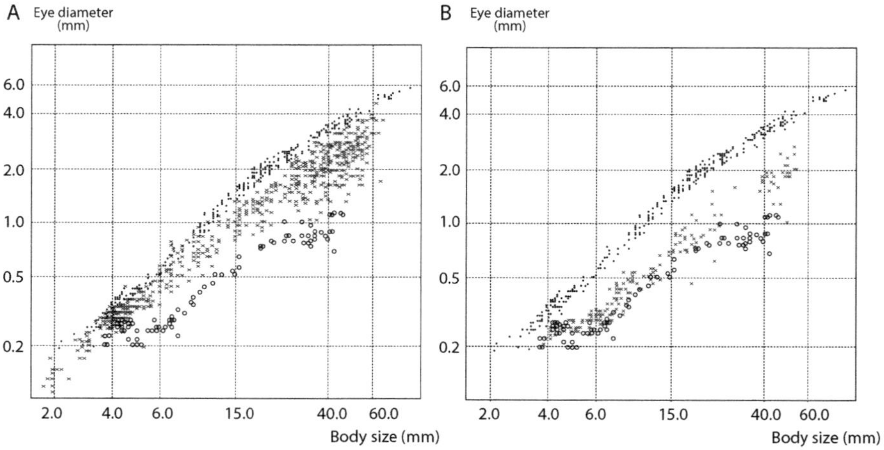


**Fig. 6.44** Relative eye growth (double logarithmic) in the F2 and the backcrossing between the *Astyanax* surface and the Pachón cave fish. (**a**) Surface fish (*dots*), F2 crossing (*crosses*), and Pachón cave fish (*open circles*). (**b**) Surface fish (*dots*), backcrossing to cave fish (*crosses*), and Pachón cave fish (*circles*), for further data see Table 6.2 (adapted from Wilkens 2007)

F2 crossing between surface and SEP cave fish (Fig. 6.44). It is exhibited, though, in the backcrossing with the SEP cave fish. This stop is correlated with further differences. In all cave fish, holes of lytic zones, caused by apoptosis, appear in both retina and lens (Fig. 6.45a). Apoptotic processes are of general importance in cave fish eye regression and were also detected in the retina rudiment of the cave cyprinid *P. andruzzii* (Berti et al. 2001; Stemmer et al. 2015). However, whereas in this species these processes are correlated with the complete loss of the eye rudiment, in *Astyanax* cave fish, cell death processes cease during stage 3 (≥6.0 mm to <15.0 mm bs) in the retina, which, along with the whole eye ball, increases in size through further growth (Figs. 6.43 and 6.44). Thus, eye growth does not arrest as recently claimed again by Krishnan and Rohner (2017).

Now, except for visual cells including outer nuclear and plexiform layers, all layers incorporating fine structures such as horizontal and Müller cells as well as the outer limiting membrane are developed in the SEP cave fish (Fig. 6.46a). Extremely rarely, single ellipsoids containing low numbers of mitochondria but no outer segments were found. The optic nerve, which is connected with the tectum opticum, as well as the falciform process are developed. The pigmentary epithelial cells have a cubic form in the central retina with their nuclei at the cell basis. The lens stagnates in its development and does not differentiate further. It consists of a capsule enclosing a noncrystalline nucleus of undifferentiated cells. In spite of only forming a rudimentary lens, the VEP Molino cave fish develops a large number of visual cells equipped with ellipsoids rich in mitochondria and outer segments at this stage (Fig. 6.46b). The lamellar arrangement of the discs of the outer segments of this population is completely regular and undisturbed (Fig. 6.47).

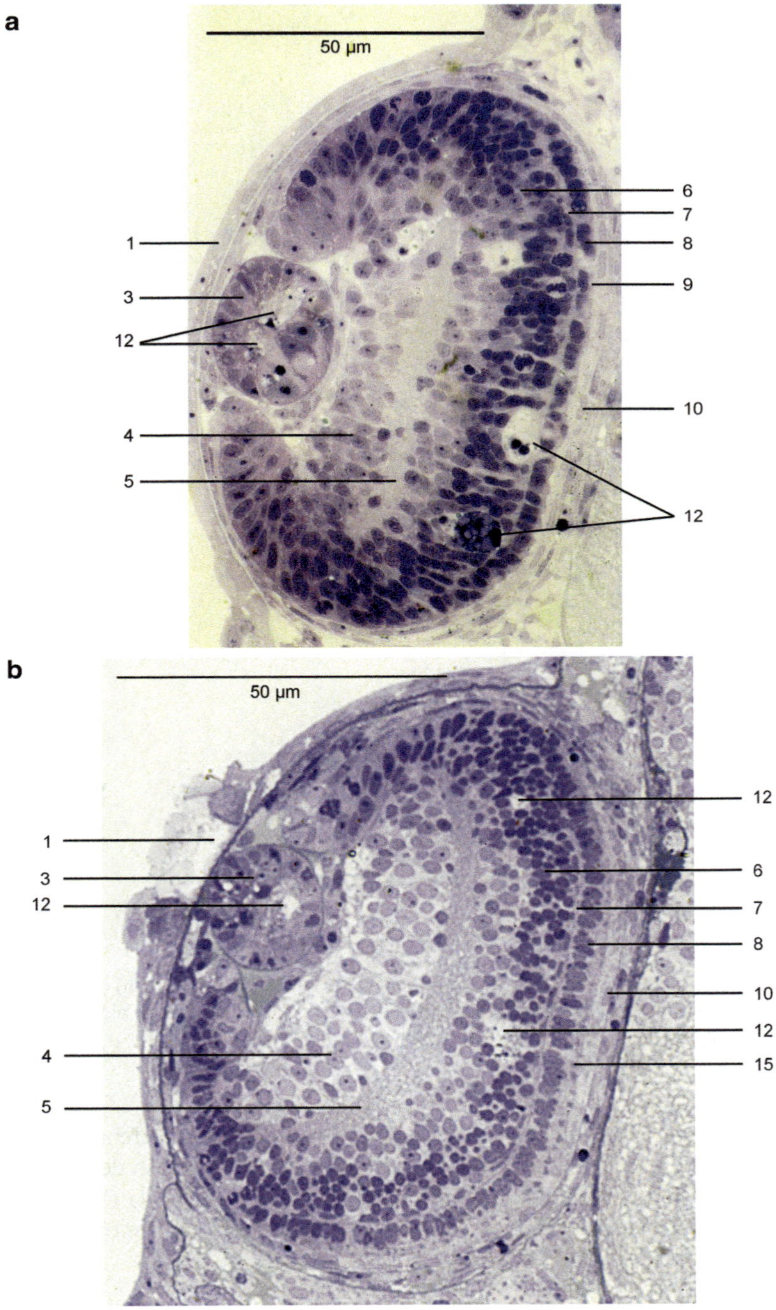


**Fig. 6.45** Semi-thin sections of the larval eye at stage 2 (5 mm body length) of the SEP Curva (**a**) and the VEP Molino (**b**) cave fish. 1 = cornea; 2 = anterior eye chamber; 3 = lens; 4 = ganglionic layer; 5 = inner plexiform layer; 6 = inner nuclear layer; 7 = outer plexiform layer; 8 = outer nuclear layer; 9 = outer limiting membrane; 10 = pigmentary epithelium; 11 = optic nerve;

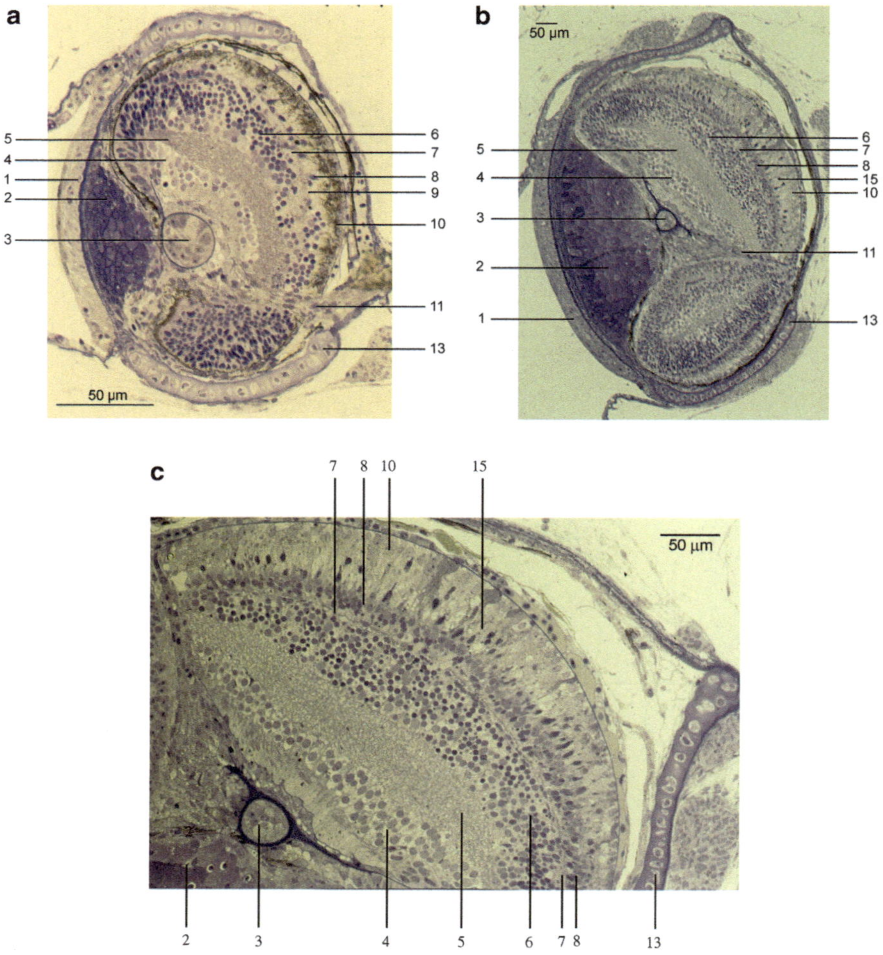


**Fig. 6.46** Semi-thin sections of the larval eye at stage 3 (12 mm body length) of the SEP Curva (**a**) and the albinotic VEP Molino (**b**, **c**) cave fish. At this stage the melanin pigmentation in the pigmentary epithelium has developed (adapted from Wilkens 2007, for abbreviations see Fig. 6.45)

During stage 4 (>15 mm bl), the degree of histological differentiation characteristic of the eye of the adult phylogenetically old SEP cave fish develops and its final process of rudimentation is reached (Fig. 6.39, 6.43, 6.44) (Wilkens 1988, 2007). The relative growth of individual eyes decreases even more to different

---

**Fig. 6.45** (continued) 12 = apoptotic centre; 13 = sclera, 15 = ellipsoids; *SEP* strongly eye- and pigment-reduced, *VEP* variably eye- and pigment-reduced (adapted from Wilkens 2007)

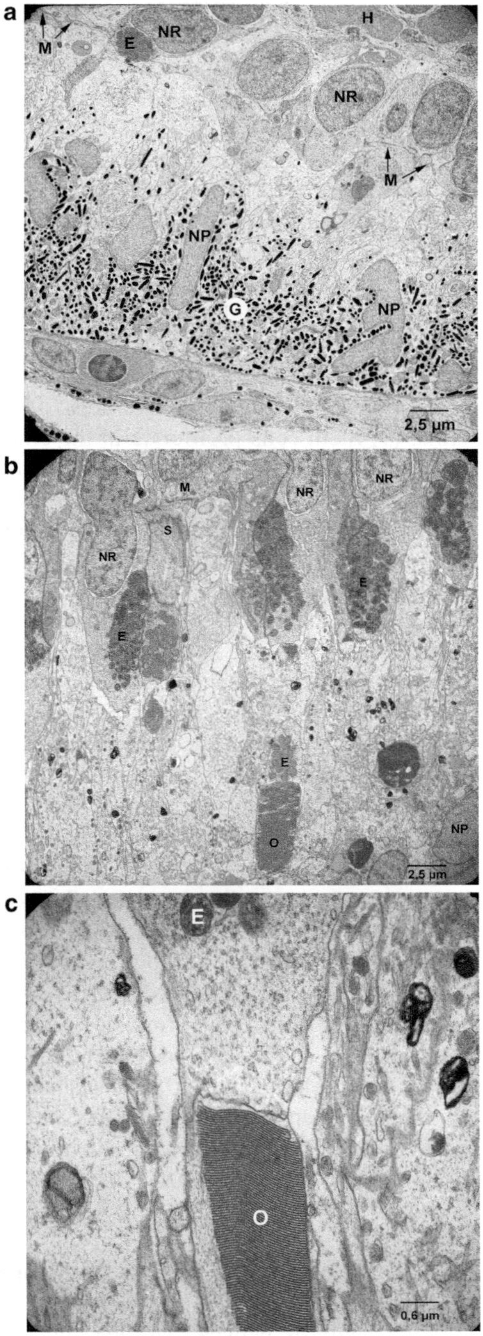


**Fig. 6.47** Ultra-thin sections of outer nuclear and pigmentary epithelial layers at ontogenetic stage 3. (**a**) Piedras cave fish (7 mm body length), (**b**) Molino cave fish (12 mm body length), (**c**) ultra-thin section of outer segment of the Molino cave fish (12 mm body length). *E* ellipsoid, *G* melanin granules in pigmentary epithelium, *H* horizontal cell nucleus, *M* outer limiting membrane, *NP* nucleus pigment epithelial cell, *NR* nucleus photoreceptor cell, *O* outer segment, *S* synaptic body (adapted from Wilkens 2007)

degrees. Due to this, the characteristic high variability of the rudimentary eye of adult cave fish develops during stage 4, but it is low and does not surpass that of the surface fish eye during stages 1–3. In the VEP Molino cave fish, the visual cells, which in contrast with the SEP cave populations were developed during stage 3, become secondarily reduced (Wilkens and Strecker 2003).

In contrast to the retina, the lens in all SEP cave fish populations will not surpass the undifferentiated histological state acquired during stage 2 and completely vanish in the last ontogenetic phase of eye reduction (stage 4). However, the VEP Molino cave population showed that visual cells with lamellar outer segments may differentiate (Fig. 6.47c), although the lens remains in the same poorly developed state characteristic of all SEP cave populations. During their ontogenetic development, the fish of this population transitorily develop retinae, which are structurally able to perceive light.

Quite generally, there are two areas of cell proliferation in the growing teleost retina. (1) Stem cells proliferate in the ciliary marginal zone of the retina (CMZ) and their progeny differentiate into various neural and glial cells except rod photoreceptors. They also provide new cells to the pigmentary epithelium (Johns 1977; Negishi et al. 1990). (2) Cell proliferation also occurs in the inner nuclear layer of the retina, where stem cells are located that give rise to rod precursors, which intercalate into the outer nuclear layer and eventually become rod photoreceptors (Kwan et al. 1996; Otteson et al. 2001). Cell proliferation is not inhibited in the CMZ of the cave fish (Strickler et al. 2007). However, whereas during stage 2 eye growth stops because it is negated by apoptotic processes in the retina, the eyes of all cave fish specimens start growing again during stage 3, although to a lesser extent than in the surface fish, as exhibited by negatively allometric growth. During stage 4, in which the variability of eye size and eye histology starts to develop, growth is only exhibited by the larger eyes whereas the smaller ones keep the same size as observed during stage 3. In summary, it can be concluded that programmed cell death (PCD) and not inhibition of cell proliferation causes the decline of retina growth in cave fish (Strickler et al. 2007). These processes almost exclusively occur during stage 2, when apoptosis develops to a large extent and the cave fish eyes temporarily stop growing (Wilkens 2007, 2010).

Also, the growth of the eye of the cave-living hepapterid catfish *R. zongolicensis* and *R. reddelli* follows the principles described for *Astyanax* cave fish. Whereas in juveniles up to about 4 cm body length the variability of eye size does not surpass that of the closely related surface sister species *R. laticauda*, it drastically increases in adult specimens and often exhibits extensive left-right asymmetries (Fig. 6.48) (Wilkens 2001). Comparable observations were made in amblyopsid cave fish (Niemiller and Poulson 2010).

### 6.21.5 Genetic Basis of Eye Development

Crossing analyses between surface and the phylogenetically old SEP cave fish revealed that eye size relies on a polygenic basis and it was calculated that in

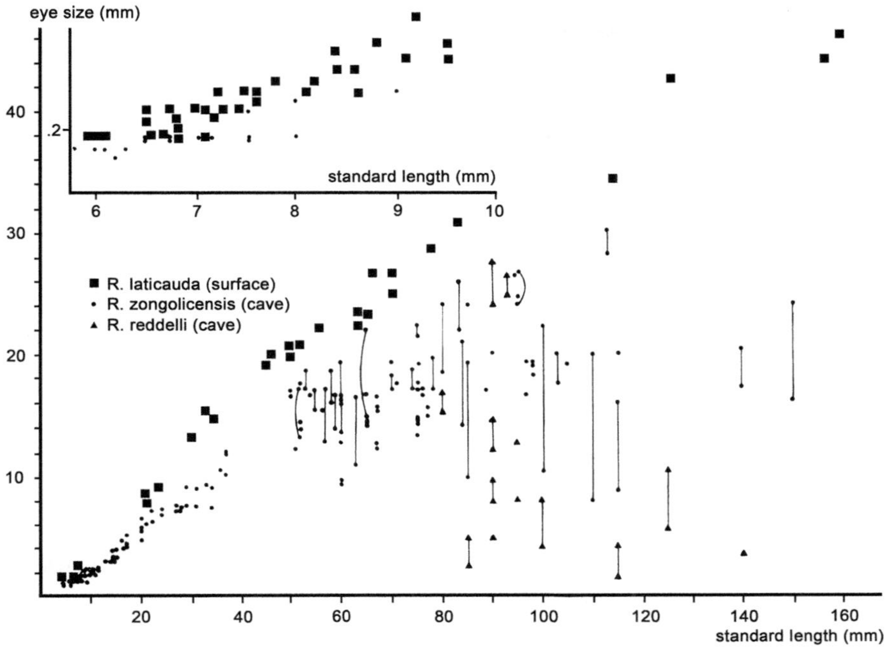


**Fig. 6.48** Ontogenetic growth curves of the eyes of the surface catfish *Rhamdia laticauda* (Hepapteridae) and the cave sister species *R. zongolicensis* and *R. reddelli*. Each *symbol* indicates eye size (measured as eye diameter) and body length (measured as standard length) of one specimen. *Connecting lines* indicate left–right asymmetry of the eye rudiments of a specific single specimen exhibiting exceptionally high divergence. Insert showing eye growth up to 10 mm body length (adapted from Wilkens 2001)

total at least eight genetic factors, the so-called "eye genes", are responsible for this difference (Wilkens 1970a, 1988; Lande 1981). The polygenic basis of eye formation was confirmed by QTL analysis (see Sect. 6.23) (Protas et al. 2007).

On the molecular level, a large number of candidate genes involved in eye development and potentially in eye reduction have been identified (Fig. 6.49) (for review see Casane and Rétaux 2016). However, in developmental control genes responsible for eye formation like *pax6*, *shh*, or *sox*, no regressive mutations have been detected in *Astyanax* cave fish (Behrens et al. 1997, 1998; Ma et al. 2014; Stemmer et al. 2015; Strickler et al. 2001).

The *Pax6* gene, which encodes a paired-class homeodomain transcription factor, is assumed to be a prime candidate for mediating eye regression in cave fish (Behrens et al. 1998; Strickler et al. 2001). In surface fish embryos, two bilateral *pax6* expression domains are present in the anterior neural plate, which extends across the midline, and fuse to form the forebrain and optic primordia. In embryos deriving from different cave fish populations, these *Pax6* domains are diminished in size and remain separated, resulting in an anterior gap in *Pax6* expression and

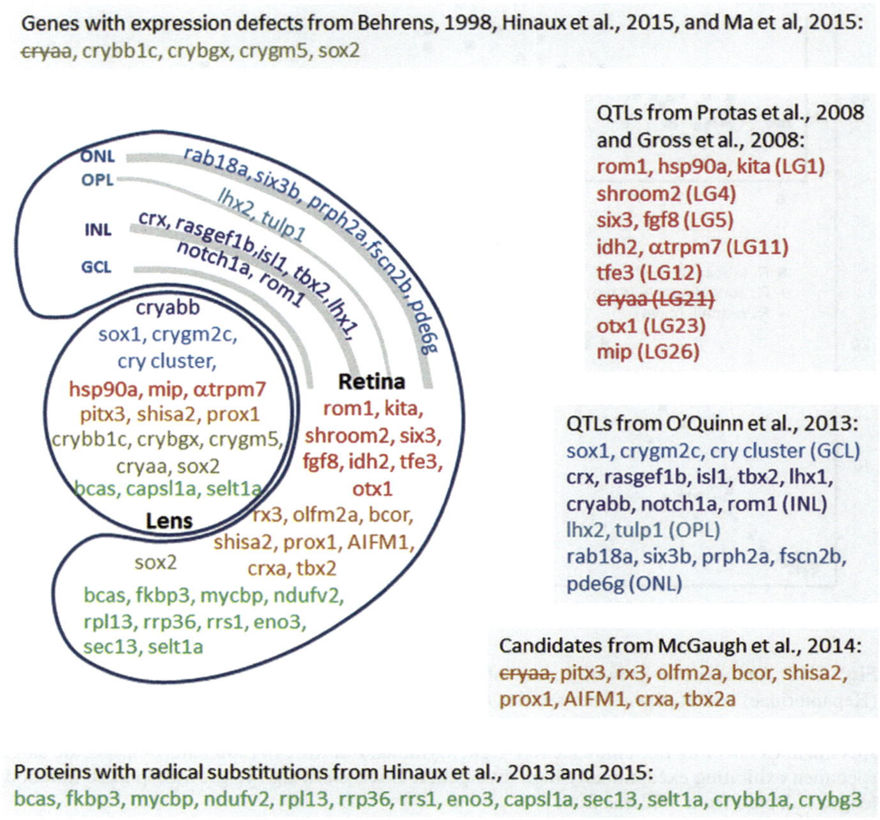


**Fig. 6.49** Genes potentially involved in eye regression of *Astyanax* cave fish indicated according to their expression patterns in the retina, the lens, or both. As *alpha-A-crystallin* (*cryaa*) expression is missing in the cavefish lens, *cryaa* is crossed in all cases because there is evidence that it is not the causal gene in the identified quantitative trait loci (QTL) interval, although absence of *cryaa* expression in the cavefish lens is deleterious to the lens. *GCL* ganglionic layer, *INL* inner nuclear layer, *ONL* outer nuclear layer, *OPL* outer plexiform layer (adapted from Casane and Rétaux 2016)

presumably the formation of smaller optic primordia (Strickler et al. 2001). *Hedgehog* proteins diffusing from the prechordal plate are known to regulate the size of the optic primordia by suppressing *Pax6* expression in the neural plate (Ekker et al. 1995; Varjosalo and Taipale 2008). Thus, the diminutive *Pax6* domains and optic primordia in cave fish embryos are suggested to be a consequence of a higher level of *hedgehog* signalling emanating from the prechordal plate. *Hedgehog* genes (*sonic hedgehog* [*shh*] and *Tiggy-winkle hedgehog* [*twhh*]) are thus obvious candidates for acting upstream of *Pax6*. As expected, the expression of these genes in several different cave fish populations was found to be expanded along the anterior embryonic midline (Yamamoto et al. 2004) and to arrest eye growth and development. These features can be phenocopied in surface fish by overexpression of *twhh* and/or *shh*, supporting a role for *hh* signalling in cave fish eye regression (Yamamoto et al. 2004; Jeffery 2005). However, mapping of candidate genes *shh*, *twhh*, and *Pax6* revealed that no eye quantitative trait loci are located near them. This result makes it unlikely that mutations in any of these genes are directly responsible for eye regression (Protas et al. 2007).

The lens plays the central role for the development of one of the two subunits responsible for eye formation (see Sect. 6.21.6) (Wilkens 2010, 2016). It was found that *alpha-A-crystallin* (*cryaa*) expression is missing in the cave fish lens (Behrens et al. 1998; Hinaux et al. 2015; Ma et al. 2014; Strickler et al. 2007). As the *cryaa* gene is located in a QTL genomic region responsible for eye loss (Gross et al. 2008; McGaugh et al. 2014), it has been proposed as a candidate for regulating cave fish lens degeneration. To reveal the underlying reason for the absence of expression in cave fish, *Sox2* was studied, which is one of the transcription factors that regulate lens crystallin genes during eye development in other species. However, *Sox2* is not expressed in the cave fish lens and it is therefore assumed that downregulation of *cryaa* in cave fish is caused by an evolutionary change in an upstream gene in the lens differentiation pathway, either *sox2* itself or a gene regulating *sox2* (Casane and Rétaux 2016; Ma et al. 2014).

Visual pigment plays a central role in the second subunit within the eye, the retina, because the disc membranes of the outer segments consist of phospholipids in which *rhodopsin* is embedded. For example, the functionality of *rhodopsin* has repeatedly been lost in different lineages of Amblyopsid cave fish (Niemiller et al. 2012) and, dependent on time and phylogenetic age, the number of retinal pseudogenes increase in eye-reduced fossorial mammals (Emerling and Springer 2014). However, knowledge regarding *Astyanax* cave fish is poor as yet, possibly because of the generally accepted prejudice of the lens being causal for promoting the regression of the eye as a whole (Wilkens 2010, 2016) (see Sect. 6.21.6). Disturbances also exist in *Astyanax* cave fish, however. In the Micos cave fish, red and green opsin gene sequences showed a high frequency of C to T changes. In Pachón, a deletion of 12 consecutive nucleotides occurs, which would disturb colour vision (Yokoyama et al. 1995). Multiple photopigments were detected in a microspectrophotometric study of the Molino cave fish (Parry et al. 2003). However, a red-like opsin was found to be still expressed in the outer nuclear layer of the Piedras cave fish for a limited period of time during early ontogeny (Langecker et al. 1993).

Gene expression and sequencing data of cave fish imply that destructive loss-of-function mutations have only occurred to a limited extent in cave fish structural eye genes. This is supported by the observation that all the different-sized eyes developed in crossings between cave and surface fish as well as in phylogenetically young VEP cave fish in principle exhibit all the structures characteristic of an eye such as lens, lens muscle, pupil, and retinal layers (Peters et al. 1975; Wilkens 1988, 2007). In crossings between strongly eye-reduced cave fish like the SEP Pachón and the VEP Molino, eyes may develop that by far outclass those of the parental cave fish by size and degree of differentiation (Fig. 6.42) (Wilkens and Strecker 2003).

Histological analyses of crossing hybrids have shown that two developmental subunits build the eye, the lens dioptric and the retina sensorial apparatus, which were suggested to inherit independently from each other, although, as in all vertebrates, the primordial eye cup induces the formation of the lens placode and both are closely correlated in size (see Sects. 6.21.3.3, 6.21.6). This apparent discrepancy can be explained by different gene systems being active during eye ontogeny. One system, which acts at early ontogeny during step 1 of eye differentiation, is responsible for the formation of a smaller eye cup (Fig. 6.50, step 1). The genes included in this group would determine its size through regulation of *hh* expression and by induction, that of the lens placode. However, in the subsequent step 2 of eye differentiation (Fig. 6.50, step 2), which includes the stages two to four of ontogenetic eye regression (Fig. 6.43), two more gene systems, one responsible for the formation of the dioptric lens subunit and another for the sensory retinal apparatus subunit, act independently from each other (see Sect. 6.21.5, Fig. 6.50).

Since the discovery that neither the *pax6*, nor the *alpha a crystalline* (*cryaa*) (Behrens et al. 1997, 1998), nor the *hedgehog* genes (Yamamoto et al. 2004) were submitted to regressive mutations, no real progress in elucidating the genetic basis of eye regression in the *Astyanax* cave fish has been made. The question of which genes regulate *shh* expression remains unanswered. In this context, the genetic study of a comparable process in snakes might provide insight. For limb regression in this group, it was shown that the ZRS enhancer of *shh* underwent a rapid increase in substitution rate from basal snakes with vestigial limbs to advanced ones without any skeletal remnants (Kvon et al. 2016). It is suggested that eye loss relies on a similar genetic basis. In that case the appearance of "back to surface eyes" in the F2

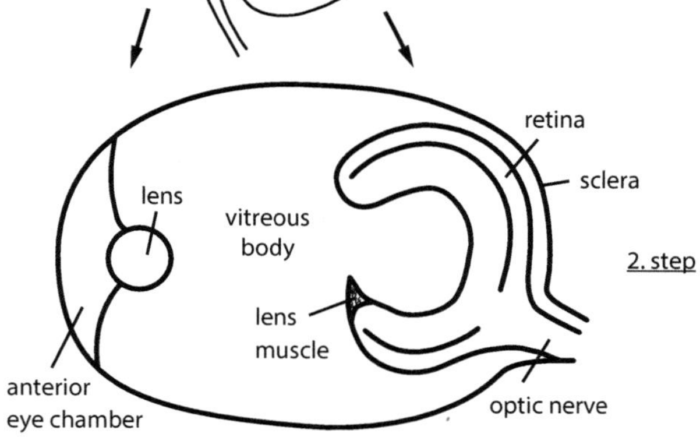


**Fig. 6.50** Scheme of the ontogenetically early first and second step of eye development in *Astyanax* cave fish (1. step and 2. step correlate to *circles* 1 and 2–4, respectively, in Fig. 6.43)

crossing between Pachón and Molino cave fish could be explained by the complementary restitution of their function brought about by the recombination of such an enhancer gene having been submitted to different mutations in geographically separate cave fish populations (Fig. 6.42) (Wilkens 1971, 2007; Wilkens and Strecker 2003).

### 6.21.6 The Role of the Lens in Eye Development

During the phylogeny of the vertebrate eye, the neuronal retina and the dioptric lens subunits have subsequently evolved (Lamb et al. 2007), which may explain their relatively large developmental independence from each other after the lens placode has been induced. This is confirmed by developmental physiological studies in chicken eyes (Coulombre 1969) and by stem cell cultures showing that the formation of the optic cup in mice occurs in the absence of a lens depending on an intrinsic self-organizing programme (Eiraku et al. 2011; Nakano et al. 2012). Like for *Astyanax*, independent development of lens and retina has also been revealed in other cave fish species like *Rhamdia* catfish (Wilkens 2001), *Sinocyclocheilus specs.* (Meng et al. 2013), and *P. andruzzi* cyprinids (Stemmer et al. 2015).

Dissenting from histological and crossing analyses in *Astyanax* that revealed that there are two independent units within the eye, a central and exclusive role of the lens for eye development as a whole in *Astyanax* was claimed (Yamamoto and Jeffery 2000; Yamamoto et al. 2003). A large number of inconclusive genetic and developmental analyses have been performed with the objective of substantiating this hypothesis (for review see Ma et al. 2014). Meanwhile, it has become a virtual axiomatic uncritically accepted fact that "Cavefish eyes are lost through apoptosis of the lens, which in turn promotes the degeneration of other optic tissues" (Casane and Rétaux 2016; Hinaux et al. 2016; Krishnan and Rohner 2017).

Based on lens transformation experiments, it was claimed that the cave fish eye can be completely "rescued" by the transplantation of a surface lens (Yamamoto and Jeffery 2000). As a proof for the asserted "rescue" of the eye, just a small number of cells expressing *rhodopsin* in the cave fish retina rudiment are provided. This finding and the conclusion drawn from this are unsubstantiated, however, because transitional expression of an opsin gene was also detected in SEP cave fish like Piedras, which never during its ontogeny forms an intact retina with visual cells (Langecker et al. 1993). Furthermore, in the reciprocal experiment in which a cave fish lens was transplanted into the surface eye or the lens vesicle was deleted from the optic cup of a surface fish embryo, the retina will develop all its characteristic layers, nonetheless. Programmed cell death and apoptosis do not occur in the surface fish larval and adult retina under these experimental conditions either (Strickler et al. 2007). Similarly, the transplantation of a surface lens into a cave fish eye did not increase the volume of the tectum opticum, in which the optic nerve fibres originating in the rescued retina would be expected to grow into. Neither did the tectum in the surface fish become thinner in specimens after excision of the embryonic lens (Rodrigues 2013).

Cave fish with transplanted surface lenses are actually developing large crystalline lenses and wide pupils (Yamamoto and Jeffery 2000). The erroneous conclusion of the complete "rescue" has probably come about by this observation. However, this finding only proves the existence of the dioptrical lens apparatus as an independent unit, as also shown by crossing experiments, but does not support the central role of the lens for retina differentiation.

Lens removal in surface fish keeps the eye smaller by about 30% (Dufton et al. 2012). As the lens influences the volume of the vitreous body the latter cannot properly develop in lens-ectomized surface fish eyes. Due to this it remains smaller in size and the eye sinks into the orbit. As a result, a cave fish eye is only superficially phenocopied. The retina, which independently develops, is malformed because of its relative oversize compared with the vitreous body, but still contains all layers including visual cell outer segments. Such retinas are still functional and even suitable for vision. Moving objects, which are sufficient for the release of aggressive behaviour in surface fish, are still recognized. This was proven by the performance of aggressive behaviour in lens-enucleated surface fish (see Fig. 1D in Espinasa et al. 2005) (see Sect. 6.12).

Also, apoptosis does not occur in lens-ectomized surface retinas. In order to explain this, the so-called "dual signal model" for retinal growth and development was secondarily introduced (Strickler et al. 2007), suggesting that the lens acts in concert with another unknown component, possibly the pigmentary epithelium. As a result, the existence of two separate units within the eye is unintentionally corroborated (Wilkens 2010, 2016). Summarizing, "complete eyes", which should contain lenses as well as retinas with fully developed visual cell outer segments, are not restored in cave fish after transplantation of a surface fish lens as is untiringly cited (e.g. Casane and Rétaux 2016; Jeffery et al. 2003; Krishnan and Rohner 2017; Yamamoto and Jeffery 2000; Yamamoto et al. 2004).

### 6.21.7 Ontogenetic Eye Regression and Head Formation

Although never achieving their genuine function, many structures that have become biologically functionless develop more or less completely during early ontogeny, but become reduced or even finally lost entirely after ontogeny has finished. Such processes are characterized as ontogenetic regression. Examples are provided by structures like the reduced pelvic girdles and hind legs of whales and sirenia, which, as small and no longer externally visible remnants, are deeply sunken into the adult body after having been almost fully developed during early ontogeny (Deimer 1977). The eyes of cave-living fish and amphibians can be quoted among such rudimentary structures, too. This was shown in species like the Cuban and Bahamian cusk eels (Bythitidae) (Eigenmann 1909; Wilkens et al. 1989), in the Mexican heptapterid catfish genus *Rhamdia* (Wilkens 2001), in the Somalian cyprinid cave fish *P. andruzzi* (Berti et al. 2001), or cave-living newts like the European cave *Proteus anguinus* (Durand 1971).

The developing eye plays a non-visual role during development as an important "organizer" of craniofacial morphogenesis (Kish et al. 2011). It directs the proper migration of cranial neural crest cells (CNCs), a transient population of migratory stem cells (Kish et al. 2011; Langenberg et al. 2008). In the developing head, retinoic acid (RA), which is synthesized in the telencephalon, eye, and nasopharynx at different times, is required for patterning. In the eye, RA is produced in a spatio-temporally regulated fashion in the dorsal and ventral fields of the retina (Matt et al. 2008). However, RA is not required for early patterning of the dorsal-ventral retina, but from the developing retina it targets RA receptors in the neural crest-derived periocular mesenchyme (Matt et al. 2008). The effects of the paracrine RA signal gradients are known to regulate important gene expression and signalling pathways. For example, altering RA signalling during early ontogeny can change a lower jaw into an upper jaw. Since RA signalling is important in craniofacial development, and since a significant amount of RA is synthesized in the retina in a tightly regulated fashion, it follows that RA from the developing eye helps to establish the RA gradient that is necessary for optimal craniofacial morphogenesis (Kish et al. 2011).

Although never achieving their genuine function, early ontogenetic anlagen may be well developed because they are responsible for the appropriate formation of other adjacent ones during early ontogeny. Consequently, they are at least transitorily still submitted to stabilizing selection. In the SEP *Astyanax* cave fish this is proven by the low variability of eye size at early ontogeny (stages 1–3), which increases drastically in adult cave fish (stage 4) (Figs. 6.43 and 6.48). The embryonic eye is supposed to play an inductive role in head and/or brain formation. In contrast, in adult cave fish, stabilizing selection for eye development relaxes and the high variability of eye ball size arises at these later stages. It is suggested that the temporary cessation of eye growth exhibited by *Astyanax* cave fish at stage 2 probably comes about by the inductive function of the eye in head and/or brain formation sinking to a lower level, because this process has been completed (see Sect. 6.21.4).

In several cave catfish, cave amblyopsids, the cave cyprinids genus *Sinocyclocheilus* (Meng et al. 2013) or cave salamanders (Weber 2000), eye reduction causes the flattening of the anterior head skeleton as a by-product.

In contrast, the outer shape of the head in *Astyanax* cave fish has not been much altered in comparison with the surface fish. Studies in F2 crossings between *Astyanax* surface and the SEP Pachón cave fish revealed that in correlation with eye size there is only a small decrease in the vertical head height of the cave fish (Wilkens, unpublished). The neurocranium is unaltered (Dufton et al. 2012) and the wide space of the orbital cavities formerly occupied by the large eye balls in surface *Astyanax* is extant in the cave fish and instead completely filled with adipose tissue (Fig. 6.1). Only the jaws are slightly elongated (Atukorala et al. 2013; Yamamoto et al. 2009).

In surface *Astyanax*, the eye is surrounded by the supra- and a semicircle of six infra- or suborbital bones, which mechanically protect it (Schemmel 1967). Number, form, and position of the orbital bones differ slightly between *Astyanax* cave populations (Dufton et al. 2012; Gross et al. 2014; Yamamoto et al. 2003) and were

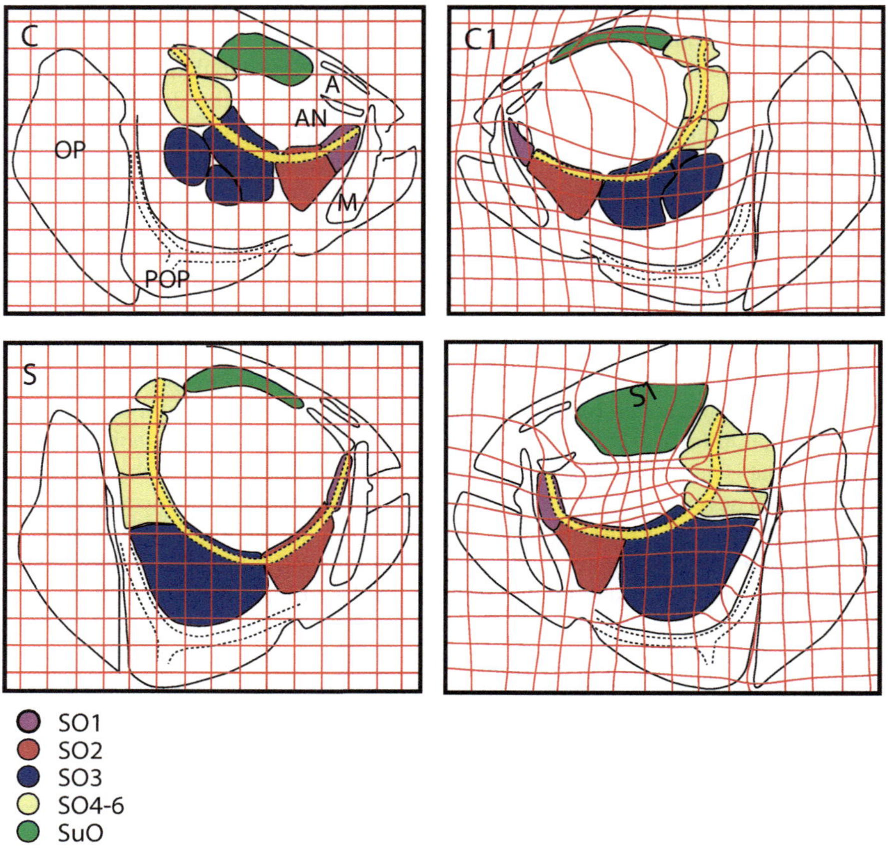


**Fig. 6.51** Craniofacial skeletons and the effects of early lens transplantation (1 day post-fertilization): Pachón cave fish (*C*) and Pachón cave fish with transplanted surface lens (*C1*), surface fish (*S*) and surface fish with transplanted cave fish lens (*S1*), *SuO* supraorbital bone, *OP* opercular, *POP* preopercular, *M* maxillary bone, *A* nasal bone, *AN* antorbital bone, *SO1–SO6* sub- or infraorbital bones (adapted from Yamamoto et al. 2003)

used for taxonomic purpose (Fig. 6.51) (Alvarez 1946, 1947; Mitchell et al. 1977). These bones also enclose the sensorially important supra- and sub- or infraorbital lateral line canals. Lentectomy and grafting of the embryonic eye between surface and cave fish and vice versa revealed the extent to which craniofacial phenotypes were influenced by experimental removal of the eye or the lens. Certain traits were affected, including the distance between the nasal and antorbital bones, the inner sectors of the third suborbital bone (SO3), the supraorbital bone, and the position of SO3 relative to the orbit of the eye. However, other craniofacial traits were not affected by eye loss, such as number of suborbital bony elements, positioning of the suborbital bones SO4 to SO6 relative to the opercular bone, and opercular bone shape (Yamamoto et al. 2003). Very often the infraorbital canal in the *Astyanax* cave fish is interrupted and shortened, because the infraorbital bones are fragmented and move in the direction of the orbit (Schemmel 1967). However, this fragmentation may be due to an artefact resulting from injuries, as was also erroneously observed for the lateral line (Wilkens 1977).

However, more severe defects in head formation like lack of the dorsal orbit and most of the ethmoid bone as observed in certain eyeless zebrafish mutants (Langenberg et al. 2008) were not developed in the experiments dealing with *Astyanax* cave fish. This may be due to the fact that an important player in craniofacial patterning was still functioning. The retina, which produces RA targeting RA receptors in the neural crest-derived periocular mesenchyme (Matt et al. 2008), is still present in the lens-ectomized surface specimens treated in these experiments and may prevent drastic cranial deformation (Wilkens 2010).

In nature, specimens exhibiting severe defects of the cranium, which, for example, might negatively impact feeding, have never been found in *Astyanax* cave fish because the specimens would have been inviable and would therefore have immediately been eliminated by selection. In contrast, minor defects observed in the *Astyanax* cave fish head skeleton, like fragmentation of the largest circumorbital bone (SO3) or partial fusion of the postorbital bones (SO4 to SO6), resulting from eye regression, do not disturb the biological function and would persist (Gross et al. 2014). Also, lateral asymmetries of the cave fish skull resulting from premature bony fusions are biologically irrelevant and therefore not eliminated.

### 6.21.8 Root Effect

The root effect is characteristic of ray-finned fishes. It is important for the supply of the retina with oxygen and for the inflation of gas into the swim bladder. Its main specificity relies on a pH-dependent reduction in hemoglobin-oxygen-carrying capacity. The underlying physiology of oxygen secretion involves an elaborate vascular arrangement, the rete mirabile, which builds the chorioidal gland in the eye. It is situated between the sclera and the retina and releases oxygen into the tissue. The choroid rete evolved only once, some 250 million years ago. In contrast, the rete mirabile of the swim bladder developed independently in different systematic groups and were secondarily lost again several times (Berenbrink et al. 2005; Rummer et al. 2013).

The chorioid in the eye is well developed in *Astyanax* surface fish but reduced in the cave fish (Figs. 6.39 and 6.41). The magnitude of the root effect has been shown to be lowered to the same degree in the VEP Micos and the SEP Pachón cave fish in comparison with the surface fish, although the size of the chorioid is variable in the Micos and much smaller in the Pachón cave fish. This indicates that the root effect magnitude inherits independently from chorioid size (Damsgaard 2016).

## 6.22 Brain

Studies of mammalian brains have revealed that the different regions develop independently from each other, which is in agreement with the "Mosaic Brain Evolution" hypothesis (Barton and Harvey 2000; Crish et al. 2006). Similar observations can be made when analyzing the specific brain parts of *Astyanax* cave fish, which also mirror sensory and behavioural adaptations to the environment differing from the surface fish. The study of 5-day-old larval and adult specimens revealed that all typical brain parts can be identified and that the cranial nerves are topologically conserved (Riedel 1997; Rodrigues 2013) (Fig. 6.52).

The most obvious finding is that the paired tecta optica along with the optic nerves are smaller in the VEP Micos and Chica as well as in the SEP Pachón cave fish in comparison with the surface fish (Figs. 6.53 and 6.54). Differences were also observed between surface and cave fish like the amblyopsids (Niemiller and Poulson 2010; Poulson 1963) or *Poecilia mexicana* (Eifert et al. 2014). In the SEP Pachón cave fish, the difference amounts were between about 50% and 65%, respectively, of the surface tectum opticum superficial area and volume (Peters

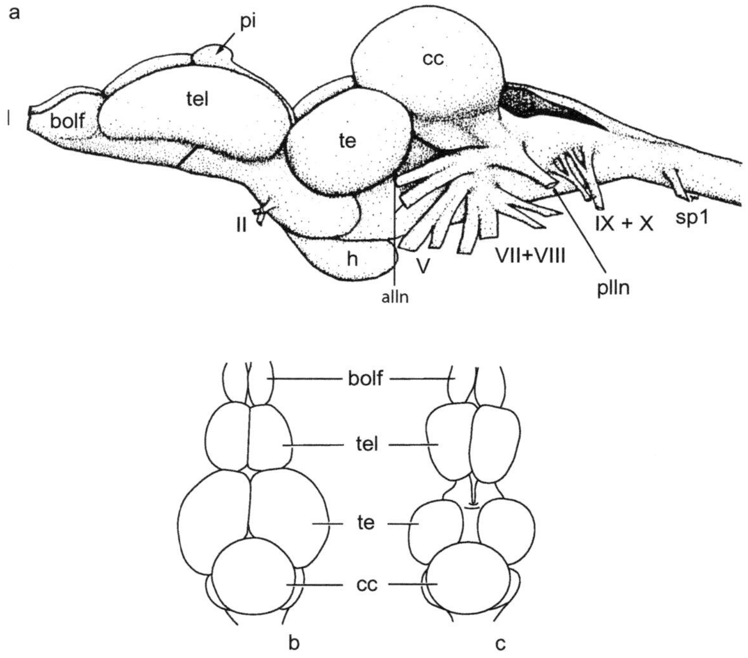


**Fig. 6.52** Lateral and dorsal view of the brain of *Astyanax* cave (**a**, **c**) and surface fish (**b**). *alln* anterior lateral line nerve, *bolf* bulbus olfactorius, *cc* cerebellum, *h* hypothalamus, *pi* pineal organ, *sp1* first spinal nerve, *tel* telencephalon, *te* tectum opticum, *plln* posterior lateral line nerve; *I* Nervus olfactorius, *II* N. opticus, *V* N. trigeminus, *VII* N. facialis, *VIII* N. vestibocochlearis, *IX* N. glossopharyngeus, *X* N. vagus (adapted from Peters et al. 1993 and Riedel 1997)

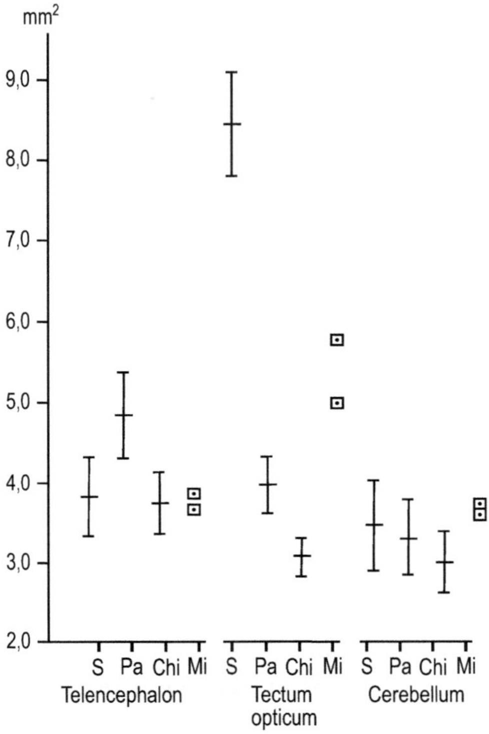


**Fig. 6.53** Mean and range of dorsal area size of telencephalon, tectum opticum, and cerebellum in 15 specimens of each of the following: *Astyanax* surface (*S*), the SEP Pachón (*Pa*), as well as the VEP Chica (*Chi*) and two Micos cave fish (*Mi*). *SEP* strongly eye- and pigment-reduced, *VEP* variably eye- and pigment-reduced (adapted from Peters et al. 1993)

et al. 1993; Rodrigues 2013), whereas in correlation to the larger eye size in the VEP Micos and Chica cave fish, the reduction is less and depends on the eye size of the specific specimens studied (Fig. 6.53) (Peters et al. 1993; Moran et al. 2015). In the SEP Pachón cave fish, the lowered volume of the tectum is already evident as early as at 5 days post-fertilization (Rodrigues 2013). This can be explained by the fact that visual cells are never developed and therefore optic innervation of the tecta does not take place. Consequently, compared with the surface fish, the tecta optica of *Astyanax* cave fish are separated by broad gaps between each other as well as from the telencephalon from the beginning of ontogeny due to their smaller size.

The olfactory bulbs are involved in the reception of olfactory input, which is then relayed to the telencephalic lobes. No differences between cave and surface fish were found in the volume of the lobes or in the number of olfactory lamellae in the naris. Therefore, it is usually assumed that the olfactory sense is not particularly enhanced in the cave fish (Riedel 1997; Schemmel 1967). However, Rodrigues (2013) observed that in contrast to the decrease in volume of the whole brain, the olfactory bulbs remain unmodified in the Pachón cave compared with the surface fish. It was therefore suggested that the olfactory bulbs are more likely enhanced (Fig. 6.54). Also, in the amblyopsid cave fish (Niemiller and Poulson 2010) and in

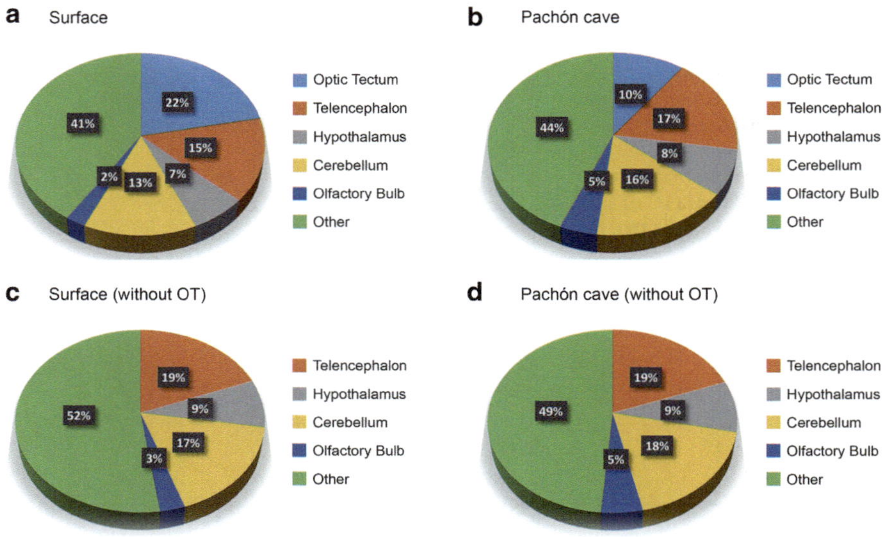


**Fig. 6.54** Ratios of the volume of different brain parts in the *Astyanax* surface fish and the SEP Pachón cave fish including (**a** and **b**) and excluding (**c** and **d**) the optic tectum. No differences except for the olfactory bulbs exist. *SEP* strongly eye- and pigment-reduced (adapted from Rodrigues 2013)

cave populations of *Poecilia mexicana* (Eifert et al. 2014), an increase of the olfactory bulbs occurs.

However, in general when excluding the tectum from that of the whole brain in surface and Pachón cave fish by calculating the percentage volume size of the different brain regions, no remarkable differences were found (Fig. 6.54) (Rodrigues 2013). Nor did Rodrigues (2013) detect a proliferation of the hypothalamus as claimed by Rétaux et al. (2008).

Also, no increase in cerebellum size was found, although this could have been expected, given its role in spatial cognition and somatosensory input from the improved lateral line system in the cave fish (Fig. 6.52). Such increase occurs in the amblyopsid cave fish as well as in *Poecilia mexicana* cave forms (Eifert et al. 2014; Niemiller and Poulson 2010; Peters et al. 1993; Rodrigues 2013). However, the anterior lateral line nerves show an enhanced diameter in the Pachón cave fish, which correlates with the improvement of the lateral line sense by number and size of neuromasts in the head region (see Sect. 6.5). In contrast, no differences were found between the surface and Pachón cave fish posterior lateral line nerves, which is in accordance with the lateral line organ not being improved along the body.

When excluding the volume of the tectum opticum, larval as well as adult Pachón cave fish exhibit a brain volume that is approximately 20% smaller than in the surface fish. It was hypothesized that this may be related to energy consumption to save high metabolic costs caused by neural tissue and information (Rodrigues 2013). It was not considered, however, that this might result from environmental modification. For example, for wild and common garden-reared surface and cave populations of *Poecilia mexicana*, a high amount of phenotypic plasticity of brain size was revealed (Eifert et al. 2014). This can not be excluded for *Astyanax* cave fish either.

## 6.23 Comparison of the Genetics of Complex Regressive and Constructive Traits

### 6.23.1 Phenotypic Manifestation and Gene Expression

The morphological and histological study of complex traits of surface and cave forms has revealed that the difference between them has quantitative character. This was shown for constructive traits like number of taste organs (Schemmel 1974a, b), feeding posture (Schemmel 1980; Kowalko et al. 2013b), and neuromasts (Wilkens, unpublished) as well as sleep duration (Duboué et al. 2011), amount of yolk content (Hüppop and Wilkens 1991), or improved ability to store fat (Hüppop 1989) and in regressive traits like eye size, melanophore number (Wilkens 1970a, b, 1988, 2010), degree of dark preference, schooling (Kowalko et al. 2013a), inclination of dorsal light reaction (Langecker 1993), sex determination, or scleral ossicles (O'Quin et al. 2015). In detail, in fish, eye regression starts with a slight decrease of total size including that of all the single structures like lens, vitreous body, or retina. Destructive histological processes can only be observed after the overall size has been considerably diminished. Loss of single structures like lens or visual outer segments only occurs in very small eyes at the end of the process of reduction (Figs. 3.17 and 6.38). A similar process was shown to take place in the decapod compound eye. At the beginning of reduction just the number of ommatidia is lower, although no structural defects are manifested. In close correlation to this, the optic ganglia or optic neuropiles (Elofsson and Dahl 1970) in the eye stalk become smaller, too. The end of this process is reached when only single or no ommatidia at all are left and also the optic ganglia are lost (Fig. 3.12).

All of these morphological, behavioural, or physiological traits differing between cave and surface fish studied in classical crossing experiments and by QTL analyses to date exhibit polygenic inheritance (Kowalko et al. 2013a, b; Protas et al. 2007; Wilkens 1988). In the following examples, the genetic basis of two regressive and three constructive traits are presented in detail.

#### 6.23.1.1 Eye

It was found that the eye size in the F2 and backcross hybrids lies between those of the respective parental forms (Fig. 6.55a). The mean of the F1 progeny is about intermediate and only slightly shifted towards the surface fish. Its distribution exhibits a much lower range than the F2 and the backcrosses. Whereas all crossings exhibit Gaussian distribution, it is bimodal in the backcross to the surface parent (Wilkens 1970a, 1988). Based on classical crossing experiments, at least eight loci, so-called "eye genes", were calculated to be responsible for eye size (Lande 1981;

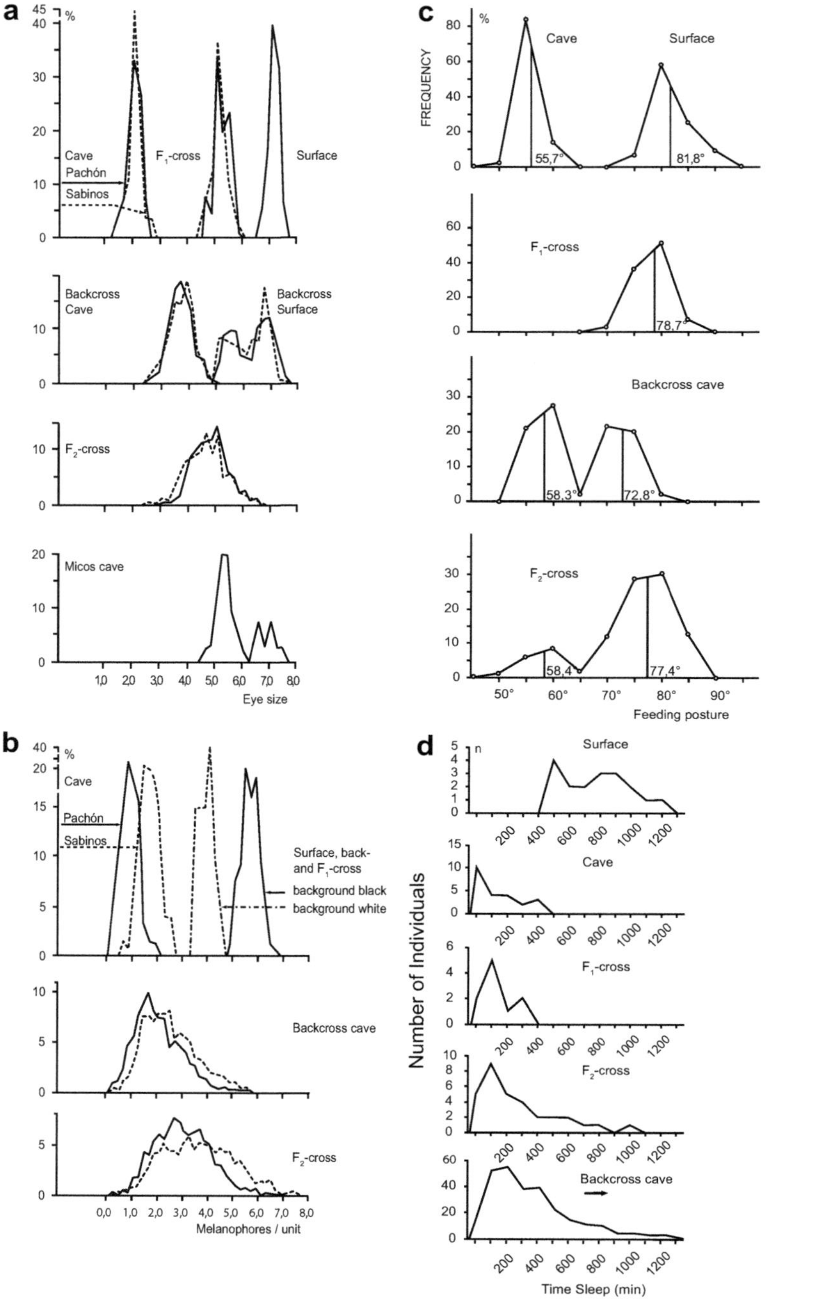


**Fig. 6.55** Frequency distributions of regressive and constructive traits in the crossings between *Astyanax* surface and cave fish. (**a**) Eye size, (**b**) melanophore number (crossings of the Sabinos and the Pachón cave fish in *dotted and continuous lines*, respectively), (**c**) feeding posture, (**d**) sleep duration, (**e**) neuromast number, (**f**) taste organ number (adapted from Wilkens 1970a, b; Schemmel 1974a, b; Duboué et al. 2011)

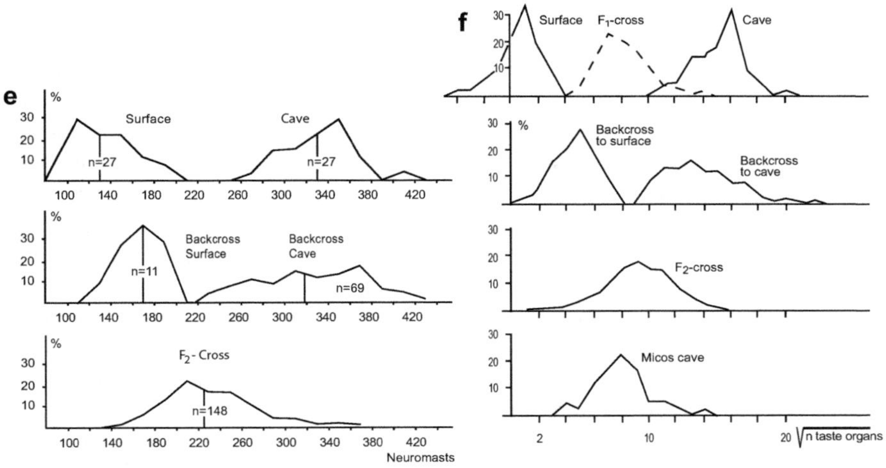


**Fig. 6.55** (continued)

Wilkens 1970a). In QTL studies it was found that 14 in total were affecting eye and lens size (Protas et al. 2007).

### 6.23.1.2 Melanophore Number

Besides the amount of melanin within the melanophores ("brown gene", see Sect. 6.20.2), the colouration of *Astyanax* surface and cave fish is also influenced by melanophore number. Mean and range do not deviate from the surface fish in the F1 and the backcross to the surface fish (Fig. 6.55b). In contrast, however, the number of melanophores developed in the F2 progeny ranges between that in the cave fish and surface fish, which are kept under a black background. Whereas across F2 progeny it is normally distributed, the progeny of the backcross to the cave fish shows positive skewness with a tail extending as far as the surface fish (Wilkens 1970b). The difference in number of melanophores found between the parental SEP Sabinos and Pachón cave fish is manifested in the F2 and the backcross to the respective cave fish population, too (Wilkens 1970b). Because of their ability to perform the morphological colour change, the surface fish only develop half the number of melanophores under a white background than under a black background (see Sect. 6.20.1). This ability was found to manifest in the F1 progeny between surface and cave fish and in the backcross to surface fish as well. In contrast, the F2 and the backcross to the cave fish do not exhibit morphological colour change, but exhibit more or less the same distributions of melanophores over a dark background as they do over a white background (Wilkens 1970b, 1988). In classical crossing experiments, at least one or two loci ("melanophore genes") were calculated to be responsible for the difference between the number of melanophores in the SEP Sabinos and Pachón cave populations, respectively. QTL mapping revealed at least 18 affecting pigmentation (Protas et al. 2007).

#### 6.23.1.3 Feeding Posture

The SEP cave fish have improved food search ability by modifying the feeding posture exhibited while taking up food from the ground (see Sect. 6.8). The classical crossing analysis revealed that the F1 progeny between surface and different phylogenetically old SEP cave fish populations exhibit similar postures to the surface fish phenotypes. The F2 progeny ranges between surface and cave fish, but the distribution is centred near the mean angle observed in the surface fish with the tails of the distribution overlapping the cave fish phenotype (Pachón and Tinaja, Kowalko et al. 2013b) or shows bimodality (Pachón and Sabinos, Schemmel 1980) with a second smaller peak weighted toward the cave fish (Fig. 6.55c). The progeny from the backcross to the cave fish exhibits bimodality with two peaks, one approximately corresponding with the cave fish and the other one partially overlapping the F1 cross. The backcross to the surface fish was not performed but is assumed to resemble the F1 cross.

Based on classical analysis, at least three loci were calculated as being involved in this difference ("feeding posture genes") (Schemmel 1980). QTL analysis confirms that the evolution of feeding posture is controlled by multiple genetic loci (Kowalko et al. 2013b). Some of them may not be shared between different cave fish populations as, for example, was revealed for the SEP Pachón and Tinaja cave fish. Consistent with these data, hybrid individuals in an F1 cross between Pachón and Tinaja cave fish have an intermediate phenotype, significantly different from the Pachón parental population. Thus, it appears that distinct QTL-representing loci are regulating feeding posture in the Pachón and Tinaja cave fish populations (Kowalko et al. 2013b). This difference was not found by classical crossing analysis between the Pachón and the Sabinos population cave fish (Schemmel 1980).

#### 6.23.1.4 Number of Free Neuromasts

For orientation and improved food finding in darkness, the number of free neuromasts is enhanced in the cave fish (Teyke 1990; Yoshizawa et al. 2012; unpublished own results). The means of the F1 and the F2 cross between Pachón cave and the surface fish are intermediate. In the F2 the distribution ranges between the cave and the surface fish with a tail extending towards the cave fish (Fig. 6.55e) (Kowalko et al. 2013b; Wilkens 2016). The mean number of neuromasts in the backcross to the surface fish is only slightly lower than in the surface fish itself, because the distribution is weighted in those surface fish exhibiting a higher number of neuromasts. The neuromasts in the backcross to the cave fish exhibit negative skewness and show a distribution reaching toward the cave fish.

#### 6.23.1.5 Number of Taste Organs

The number of taste sense organs developed below the ventral jaw is increased in the SEP cave fish compared with the surface fish (Fig. 6.55f) and was found to cover a larger area containing more organs in the Pachón than in the Sabinos cave fish. However, this difference between the two cave fish populations is not manifested in their crossings (Schemmel 1974a). In the SEP Sabinos cave fish the means of the backcrossing to the cave fish and of the F2 crossing between surface and SEP cave fish are shifted towards the cave fish. The numbers of taste organs exhibit Gaussian distributions ranging between the means of the respective parental forms. However, the backcross to the cave fish shows a broader distribution overlapping almost entirely with the cave fish. Three loci were calculated by classical crossing analysis and by QTL studies to be responsible for the enhancement of the taste organs ("taste bud genes") (Protas et al. 2007). Their number in the phylogenetically young VEP Micos cave fish ranges between the means of the surface and the Sabinos cave fish with a tail extending to the latter (Schemmel 1974b).

#### 6.23.1.6 Sleep Duration

Change in ecological conditions, from surface to cave, was found to be correlated with a change in sleep duration and activity, which was revealed by studies of three independently derived *Astyanax* cave populations, the SEP Tinaja, Pachón, and the VEP Molino cave fish (see Sect. 6.10, Duboué et al. 2011). Classical crossings were performed between surface and Pachón cave fish. It was found that the range of the F1 crossing is congruent to the cave fish. In the F2 and the backcrossing to the cave fish, a peak is weighted in the cave fish with a tail overlapping the surface fish (Fig. 6.55d). QTL analysis showed that multiple loci ("sleep genes") are involved in the difference of sleep duration.

### 6.23.2 Comparison of Phenotypic Manifestation

Classical crossing experiments revealed that the complex traits studied in *Astyanax* surface and cave populations rely on polygenic systems. This was supported by QTL analyses, in which the number of genes was usually higher (e.g. 8 vs 14 for eye plus lens size, 4 vs 18 affecting dark pigmentation) (Protas et al. 2007). This can probably be attributed to the fact that groups of genes involved may be located on one chromosome, which cannot be differentiated by classical crossings.

The classical crossings of *Astyanax* do not exhibit fundamental differences in the phenotypic manifestation between constructive and regressive traits, but are submitted to principally comparable patterns of distribution and phenotypic manifestation. They exhibit an exponentially rising epistatic gene effect, the amount of which differs in the various traits. The contribution of gene effect may continuously increase in an "asymmetrical" manner towards the specific larger parental trait of the surface (e.g. eye size or melanophore number) or the cave fish (e.g. taste bud or neuromast numbers). When just skewness of distribution curves is observed, epistatic gene effect size is only slightly enhanced, building a distributional tail (e.g. Fig. 6.55b, backcross to cave; Fig. 6.55d, F2 and backcrossing to cave). As a result, the range is considerably enhanced in these crossings. In contrast, bimodality is caused when, by an abrupt "threshold-like" increase of effect, a second additional distribution peak appears, which converges with the distribution of the larger traits (e.g. Fig. 6.55a, backcross to surface and Micos cave fish; Fig. 6.55c, backcross to cave and F2 crossing). Finally, complete "dominance" due to large epistatic gene effect is observed. In this third case, the critical epistatic "genic threshold" of gene effect is so large that the intermediate traits do not build one of the two peaks found in the stage presented before (e.g. Fig. 6.55b, F1 and backcrossing to surface; Fig. 6.55c, F1 crossing; Fig. 6.55d, F1 crossing). By these epistatic gene effects the means of most traits are shifted towards the larger trait. The "threshold" might imply that a monogenic mode of inheritance would be responsible for this. This was rejected by crossings, however, as shown for feeding posture (Schemmel 1980).

However, in some traits the pattern of exponential asymmetrical increase of gene effect unexpectedly does not manifest in the F2 crosses between surface and cave fish. For example, trait sizes do not develop bimodality for eye size as expected with respect to the distribution curve in the backcross to the surface fish (Fig. 6.55a). Neither can this be observed in the F2 crossing for melanophores, although the F1 cross converges with that of the surface fish (Fig. 6.55b). For both traits, nearly Gaussian distribution is exhibited in these F2 crosses. The latter cannot be explained by the loss of gene balance as might be the expected result from the recombination of two genetically separated populations, because, for example, other traits like feeding posture exhibiting bimodality (Fig. 6.55c) or sleep durance skewness (Fig. 6.55d) develop enhanced gene effect in the respective F2 crosses. I suggest that the manifestation of epistatic gene effect in these F2 crosses is inhibited, because both eye size and melanophore number are based on two independent subunits. These are in the eye lens and retina subunits (see Sect. 6.21.3.1) and in pigmentation melanophore number and the ability of morphological colour change (Sect. 6.23.1.2). The alleles of the different genes responsible for the different subunits are recombined at random (Wilkens 2010). In addition to eye and melanophores, Gaussian distribution is also found for taste organs in the F2 crossing between SEP cave and surface fish. It is suggested that this might also be caused by the involvement of two independently inheriting traits, namely taste buds and solitary chemosensory cells (SCCs), which were incorrectly counted as the same organ by Schemmel (1967) (Fig. 6.55f).

In case of the eye, a series of findings have corroborated the existence of two developmental units. For example, histological analyses revealed that two units built by the dioptric lens apparatus and the sensory retina apparatus form the eye (see Sect. 6.21.3.3, Wilkens 2010). Further support is provided by the lens size of the F2 progeny still exhibiting bimodality, when measured separately, although the eye ball size shows Gaussian distribution (Wilkens 1970a). As concerns the retina, it is difficult to detect and quantify such deviating phenotypic manifestations. However, studies of the surface fish, the SEP Pachón, and the VEP Micos cave fish revealed that brain and eye mass are closely correlated in these forms (see Sect. 6.22) (Moran et al. 2015; Peters et al. 1993; Pfeiffer 1967). In contrast, the brain mass is largely invariant despite a ten-fold difference in eye mass in the F2 crossing between surface and Pachón cave fish, and no clear correlation of external eye size and brain tectum size as expected from the findings in pure cave fish was detected (Moran et al. 2015). As tectum opticum mass directly depends on the amount of retinal nerve cells, this result indicates that like the lens, the retina also shows a more independent manner of manifestation from that of the eye ball in the variable

F2 crossings. Furthermore, the VEP Molino cave fish, which never develops a differentiated lens, but nonetheless completely independently develops complete visual cells containing outer segments at early ontogeny, substantiates this (Wilkens 2007). The existence of two independent subunits within the eye is also supported by the so-called "dual signal model" (Strickler et al. 2007) for retinal growth and development, which suggests that the lens acts in concert with another unknown component and it is hypothesized that this would be the pigment epithelium, which is part of the second retinal, but not of the lens subunit.

Two separate units are also responsible for the melanophore system. In addition to the formation of melanophores, the number developed on the body surface is influenced by a second independent unit, the morphological colour change over different coloured backgrounds. As mentioned before, this ability to change colour is lost in the SEP cave fish (see Sect. 6.20.1). Both the formation of the melanophores and the morphological colour change traits are influenced by a large number of genes (Protas et al. 2007; Wilkens 1970b, 1988). Therefore, due to their independent recombination and assortment, the epistatic effects, which are responsible for the functioning of the morphological colour change, also do not manifest.

In addition, the Gaussian distribution of the number of taste organs found by Schemmel (1974a, b) in the F2 crossing between SEP cave and surface fish is probably caused by the disturbed manifestation of epistatic gene effect. This may well be explained by the involvement of two units. Schemmel (1974a) did not differentiate between real taste buds and solitary chemosensory cells (SCCs), the existence of which was unknown at that time (see Sect. 6.6) (Yamamoto et al. 2009, own observations). I argue therefore that the distribution curve originally found in the F2 crossings is based on counting both types of organs. The bimodal distributions assumed to exist for both units are supposed to be blurred because of the independent inheritance.

### 6.23.3 Genetics of Phylogenetically Young VEP Cave Fish

The most obvious feature of the VEP populations Micos, Chica, and Caballo Moro is that their externally visible eyes exhibit variable size and are less reduced than those of the SEP cave fish. As mentioned above, these cave fish populations have long been thought to have a more recent origin or are just hybrids between the surface and SEP cave fish (Bradic et al. 2012; Mitchell et al. 1977; Strecker et al. 2012; Wilkens 1988; Wilkens and Burns 1972).

Classical genetic analyses of the VEP Micos cave fish revealed that the inheritance of regressive and constructive traits in this fish itself, as well as in its crossings with the SEP cave fish and the surface fish, is in principle subjected to the same rules of manifestation, as they were analyzed by crossing SEP cave fish and the surface fish (Wilkens 1976).

The melanophore number of the Micos cave fish kept at daylight does not differ from that of the surface fish and the morphological colour change is also still functioning. Consequently, none of the crosses of the Micos cave fish with the surface fish (Fig. 6.35, F1 (Micos × Surface), Back (Micos × Surface) × Micos, Back (Micos × Surface) × Surface) diverge from the parental surface fish and Micos cave populations. They can perform the morphological colour change and show equivalent melanophore densities. The same is valid for the F2 crossing between Micos and surface fish (Fig. 6.35, F2 Micos × Surface).

In contrast, however, to the F1 cross between the SEP cave fish (Sabinos and Pachón) and the surface fish (Fig. 6.55b), the F1 cross between the Sabinos and Pachón cave fish and the Micos cave fish does not exhibit the same number of melanophores as the surface fish, but only shows an intermediate number. Also, the melanophore number in the backcross of the F1 (Micos × Cave) to the Micos fish is more or less intermediate (Fig. 6.35, Back (Micos × Cave) × Micos). In contrast, the backcross of the F1 (Micos × Cave) to the SEP cave fish (Fig. 6.35, Back (Micos × Cave) × Cave) lies in between the F1 cross and the SEP cave fish. As can be derived from the slight difference of means, only a few of the F1 and backcross hybrids are able to perform the morphological colour change over white and black backgrounds. The melanophore number of the F2 crossing between Micos and SEP cave fish ranges between both (Fig. 6.35 = F2 Micos × Cave). The small difference of the means over black or white backgrounds in the before-mentioned crossings indicates that the morphological colour change only functions in a few of the hybrids. Although the Micos cave fish is still able to perform a morphological colour change, the previously described failures of its manifestations in the different crossings with SEP cave fish can be explained by the genetic basis of the morphological colour change in the Micos cave fish already carrying regressive mutations. Whereas the high exponential increase of gene effect necessary to develop the full number of melanophores is still manifested in the Micos cave fish, the number of non-mutated genes does not suffice to develop this epistatic increase to perform the morphological colour change in the above crossings with the SEP cave fish.

The eye size of the laboratory-bred Micos cave fish is distributed between the surface fish eye and half its size. However, in Micos cave fish "surface eyes" can only be achieved by selectively crossing specimens equipped with comparatively larger eyes (Wilkens 1976), which are rare under natural conditions (Fig. 5.7). Most remarkably, the distribution of eye size is bimodal (Figs. 5.5 and 6.55a). This indicates that in contrast to the F2 crossing between surface and SEP cave fish, enhanced epistatic gene effect can manifest in the Micos cave fish. Such enhanced epistatic gene effect is also exhibited in the distribution curve of the Micos cave fish taste organs, which exhibits skewness. Its peak correlates with that of the F1 cross between surface and SEP cave fish, but a skew distributional tail extends to the latter (Schemmel 1974b) (Fig. 6.55f). In addition, the variability of pupillary opening and lens size (which are correlated) of the Micos cave fish is lower than in its F2 crossings with the surface fish or with the SEP cave fish. It is also lower than in the F2 crossings between the surface and the SEP cave fish (Fig. 6.40).

The results of classical crossing experiments performed with the VEP Micos cave fish show that the basal principles of manifestation of regressive as well as constructive traits are the same as in the crossings between phylogenetically old

SEP cave and the surface fish. There are some differences, though, corroborating that this population is no hybrid but phylogenetically young. This is above all the undisturbed manifestation of the discontinuous increase of epistatic gene effect in eye size and melanophores as well as the low variability of the pupil size.

### 6.23.4 Significance of Unproportional Epistatic Gene Effect

The existence of VEP cave fish populations like Micos exhibiting intermediate stages of regressive traits (eyes, melanophores, dorsal light reaction, or aggressive behaviour) and constructive traits (taste organs, free neuromasts, feeding posture, egg yolk content, naris size, or fat storage ability) comparable to those of the SEP cave fish provides an exceptional tool to evaluate whether the genetic findings revealed by classical crossing experiments between surface and SEP cave fish can also be observed and above all are relevant during the natural evolutionary process (Aspiras et al. 2015; Bibliowicz et al. 2013; Hüppop and Wilkens 1991; Wilkens 1988, 2010). This could be confirmed in the Micos cave fish, in which classical genetic analyses showed that the inheritance of regressive and constructive traits is subjected to the same rules as were revealed by the crossing analyses between the surface and the phylogenetically old SEP cave fish (Wilkens 1976, 1988, 2010).

For example, the bimodal distribution of eye size observed in the VEP Micos cave fish (Fig. 6.55a) derives from a threshold-like increase of epistatic gene effect. The recombination of up to a certain number of eye genes all exhibit the same amount of small gene effect. However, after the contribution of additional eye genes, a second peak is formed because epistatic gene effect increases manifesting a threshold. This may be achieved by breeding large-eyed specimens in the laboratory. It may also happen in nature as long as there is a sufficient number of alleles present in the gene pool. The probability of such recombination depends on the number of unmutated genes left. If it is low, large eyes will only rarely appear.

In general, a quantitative trait locus may exhibit dominance, additive, non-additive, or recessive gene effect (Shao et al. 2008). The crossings between the *Astyanax* surface and cave populations have revealed that the genes responsible for the phenotypic manifestation of regressive and constructive traits in principle act in a quantitative manner. The phenotypic manifestation comes about by the additive effect of single genes. The extent of each gene effect is regulated by interaction of the whole genetic system, whereby one gene interferes with the phenotypic expression of one or more non-allelic ones. The amount of increase may be the same or be unproportional and in the extreme show discontinuous threshold-like increases. However, it is the phenotypically manifested trait that is submitted to selection and therefore the underlying epistatic effects of each single gene of the gene system as a whole are shaped by selection.

Therefore, the suggestion of a potential significance of the differing epistatic gene effects and in particular of the threshold-like one arises. In some traits, a biological function was detected. For example, the phenotypic manifestation of the number of melanophores is reversible after the total number of melanophore genes is recombined. As a result, the adaptive morphological colour change can be performed. The threshold-like epistatic gene effect obviously also plays a role in monogenic sex determination, which is acquired by the cave fish. They have switched from the additive polygenic sex determination occurring in the surface ancestor to a quasi-monogenic one in the cave fish. In *Astyanax* cave fish, monogenic sex determination results from a threshold-like increase of epistatic gene action, by which an autosomal gonosome originates (see Sect. 6.3). However, for most other traits the threshold-like epistatic gene effect is suggested to play a role in constructive evolution. With the discontinuous quantitative increase of traits by number, size, or intensity in the case of behaviour, selection is enabled to promote a phenotype more rapidly the larger the constructive phenotypical step achieved is. This is advantageous because small quantitative changes of constructive traits, as provided by small gene effect, are overruled by modificatory influence. Thus, unproportional increase of epistatic gene effect may be an important means of constructive evolution.

This finding may furthermore explain why the evolution of regressive traits in *Astyanax* cave fish is often characterized as proceeding rapidly. For example, due to the "threshold-like" decrease of gene effect at the beginning of eye reduction, its size abruptly gets smaller by about 25% of the surface fish eye. In contrast, the subsequent phenotypic regressive process of eye reduction ending in complete regression proceeds much more slowly, because gene effects are smaller.

### 6.23.5 Nature of Genes Responsible for Complex Traits

The analyses of crossing experiments suggest that the genetic expression and phenotypic manifestation of regressive traits such as the reduced eye or melanophore numbers, and of constructive traits such as feeding posture or neuromast numbers, as well as all other complex traits analyzed, are polygenic and show similar principles of manifestation. Therefore, it is suggested that the above-described findings provide insight into the evolution of complex regressive and constructive traits in general.

Because of their quantitative nature of expression, the "eye genes" or "melanophore genes" found by classical crossing experiments in regressive traits as well as "neuromast genes" or "feeding posture genes" found in constructive ones are suggested to be regulatory genes. Apart and independently from the structural genes which are responsible for each specific trait, they strongly determine the final size of a trait already at early ontogeny (Wilkens 2010). This assumption gets support from findings in eye regression. They show that in the *Astyanax* cave fish eye regression is due to expanded *hedgehog* (*hh*) gene expression along the anterior midline (Yamamoto and Jeffery 2000; Yamamoto et al. 2004, 2009). It suggests that already the very first anlage of the eye is smaller. Mapping of candidate genes *shh*, *twhh*, and *Pax6* revealed that no eye QTLs are located near these loci. This result makes it unlikely that mutations in any of these genes are directly responsible for eye regression (Protas et al. 2007). However, the question of what causes downregulation of *shh* genes and consequently determines the initial eye size remains unsolved. It is suggested that as-yet unidentified genes are regulating *shh* expression (Wilkens 2010; Yamamoto et al. 2009), which would be equivalent to those genes called "eye genes" found by crossing experiments and which are hypothesized to be responsible for eye size (Wilkens 2010). These genes would also determine the variability of the embryonic eye sizes characteristically found in the diverse crossings already at early ontogeny, because they regulate *shh* expression to different extents.

In this context, the progressive reduction of limbs from basal snake species only carrying rudimentary ones to advanced snakes in which all skeletal structures have disappeared provides an example (Kvon et al. 2016). During limb development a ZRS enhancer (Zone of Polarizing Activity Regulatory Sequence) is active in the limb bud mesenchyme where it is critically required for normal limb development in vertebrates. The ZRS is a limb-specific enhancer of the *sonic hedgehog* (*shh*) gene that is located at the extreme distance of nearly one million base pairs from its target promoter. It is highly conserved in basal snakes, whereas it underwent a rapid increase in substitution rate in advanced ones. The identification of snake-specific sequence changes within the otherwise highly conserved long range limb enhancer of *sonic hedgehog* provides a possible explanation of the complex pattern of how *sonic hedgehog* may be regulated and embryonic eye size is determined. It may also exemplify how all the other constructive and regressive traits evolve in the *Astyanax* cave fish.

It was therefore proposed that the genes responsible for eye development are divided into two groups: the first group expresses around the *shh* genes during the first step of eye regression and determine the size of the primordial eye cup through regulation of *shh* expression (Fig. 6.50) (Wilkens 2010). The second group of eye genes is suggested to regulate structural genes responsible for the lens and the retina subunits, respectively (see Sect. 6.21.4, 6.21.5).

## References


Albrecht U, Ripperger JA (2009) Clock genes. In: Binder MD, Hirokawa N, Windhorst U (eds) Encyclopedia of neuroscience, Part 3. Springer, Berlin, p 759–762

Alunni A, Menuet A, Candal E et al (2007) Developmental mechanisms for retinal degeneration in the blind cavefish *Astyanax mexicanus*. J Comp Neurol 505:221–233

Alvarez J (1946) Revisión del genero *Anoptichthys* con descripción de una especia nueva (Pisces, Characidae). An Esc Nac Cien Biol Mex 4:263–282

Alvarez J (1947) Descripción de *Anoptichthys hubbsi* caracinido ciego de la Cueva de Los Sabinos, S. L. P. Rev Soc Mex Hist Nat 8:215–219

Aspiras AC, Rohner N, Martineau B et al (2015) Melanocortin 4 receptor mutations contribute to the adaptation of cavefish to nutrient-poor conditions. Proc Natl Acad Sci USA 112:9668–9673

Atukorala AC, Hammer C, Dufton M et al (2013) Adaptive evolution of the lower jaw dentition in Mexican tetra (*Astyanax mexicanus*). EvoDevo 4:28. doi:10.1186/2041-9139-4-28

Ball EG, Jungas RL (1965) Net gas exchange and oxygen consumption. In: Renold AE, Cahill Jr (eds) Adipose tissue, Handbook of Physiology, vol 5. American Physiological Society, Washington DC, pp 355–361

Banister KE (1984) A subterranean population of *Garra barreimiae* (Teleostei: Cyprinidae) from Oman, with comments on the concept of regressive evolution. J Nat Hist 18:927–938. doi:10.1080/00222938400770811
Banister KE, Bunni MK (1980) A new blind cyprinid fish from Iraq. Bull Br Mus (Nat Hist), Zool 38:151–158
Barton RA, Harvey PH (2000) Mosaic evolution of brain structure in mammals. Nature 405:$32#1055–1058
Beale A, Guibal C, Tamai K et al (2013) Circadian rhythms in Mexican blind cavefish *Astyanax mexicanus* in the lab and in the field. Nat Commun 4:1–10. doi:10.1038/ncomms3769
Behrens M, Langecker GT, Wilkens H et al (1997) Comparative analysis of Pax-6 sequence and expression in the eye development of blind cave fish *Astyanax fasciatus* and its epigean conspecific. Mol Biol Evol 14:299–308
Behrens M, Wilkens H, Schmale H (1998) Cloning of the alphaA-crystallin genes of a blind cave form and the epigean form of *Astyanax fasciatus*: a comparative analysis of structure, expression and evolutionary conservation. Gene 216(2):319–326
Berenbrink M, Koldkjær P, Kepp O et al (2005) Evolution of oxygen secretion in fishes and the emergence of a complex physiological system. Science 307:1752–1577
Berti R, Durand JP, Becchi S (2001) Eye degeneration in the blind cave-dwelling fish *Phreatichthys andruzzi*. Can J Zool 79(7):1278–1285. doi:10.1139/z01-084
Beukeboom L, Perrin N (2014) The evolution of sex determination. University Press, Oxford, 212 pp
Bibliowicz J, Alié A, Père S et al (2013) Differences in chemosensory response between eyed and eyeless *Astyanax mexicanus* of the Rio Subterráneo cave. EvoDevo 4:25–31
Bilandžija H, Parkhurst A, Jeffery WR (2013) A potential benefit of albinism in *Astyanax* cavefish: downregulation of the oca2 gene increases tyrosine and catecholamine levels as an alternative to melanin synthesis. PLoS One 8:1–14
Bleckmann H, Mogdans J, Engelmann J et al (2004) Wie Fische Wasser fuehlen: Das Seitenliniensystem. BIUZ 34(6):358–365
Bolliet V, Ali MA, Lapointe FJ et al (1996) Rhythmic melatonin secretion in different teleost species: an in vitro study. J Comp Physiol B 165:677–683
Borowsky RL, Cohen D (2013) Genomic consequences of ecological speciation in *Astyanax* Cavefish. PLoS One 8(11):1–8. e79903. doi:10.1371/journal.pone.0079903
Boudriot F, Reutter K (2001) Ultrastructure of the taste buds in the blind cave fish *Astyanax jordani* ("*Anoptichthys*") and the sighted river fish *Astyanax mexicanus* (Teleostei, Characidae). J Comp Neurol 434:428–444
Bradic M, Beerli P, García-de León F et al (2012) Gene flow and population structure in the Mexican blind cavefish complex (*Astyanax mexicanus*). BMC Evol Biol 12:1–16
Burchards H, Dölle A, Parzefall J (1985) Aggressive behavior of an epigean population of Astyanux *mexicanus* (Characidae, Pisces) and some observations of three subterranean populations. Behav Processes 11:225–235
Burt de Perera T (2004a) Fish can encode order in their spatial map. Proc Biol Sci 271:2131–2134
Burt de Perera T (2004b) Spatial parameters encoded in the spatial map of the blind Mexican cave fish, *Astyanax fasciatus*. Anim Behav 68:291–295
Caballero-Hernandez O, Hernandez-Patricio M, Sigala-Regalado I et al (2015) Circadian rhythms and photic entrainment of swimming activity in cave dwelling fish *Astyanax mexicanus* (Actinopterygii: Characidae), from El Sotano La Tinaja, San Luis Potosi, Mexico. Biol Rhythm Res 46:579–586
Cahn PH (1958) Comparative optic development in *Astyanax mexicanus* and two of its blind cave derivatives. Bull Am Mus Nat Hist 115:69–112
Campbell SS, Tobler I (1984) Animal sleep: a review of sleep duration across phylogeny. Neurosci Biobehav Rev 8:269–300
Capellini I, Nunn CL, McNamara P et al (2008) Energetic constraints, not predation, influence the evolution of sleep patterning in mammals. Funct Ecol 22(5):847–853. doi:10.1111/j.1365-2435.2008.01449.x

Casane D, Rétaux S (2016) Evolutionary genetics of the cavefish *Astyanax mexicanus*. Adv Genet 95:117–195

Cavallari N, Frigato E, Vallone D et al (2011) A blind circadian clock in cavefish reveals that opsins mediate peripheral clock photoreception. PLoS Biol 9(9):e1001142

Chaves I, Nijman RM, Biernat MA et al (2011) The *Potorous* CPD photolyase rescues a cryptochrome-deficient mammalian circadian clock. PLoS One 6(8):e23447

Chivers DP, Wisenden BD, Hindman CJ (2007) Epidermal 'alarm substance' cells of fishes maintained by non-alarm functions: possible defence against pathogens, parasites and UVB radiation. Proc Biol Sci 274(1625):2611–2619

Coombs S, Braun CB, Donovan B (2001) The orienting response of Lake Michigan mottled sculpin is mediated by canal neuromasts. J Exp Biol 204:337–348

Coulombre AJ (1969) Regulation of ocular morphogenesis. Invest Ophthalmol Vis Sci 8:25–31

Coulombre JL, Coulombre AJ (1963) Lens development: fibre elongation and lens orientation. Science 142:1489–1490

Crish SD, Dengler-Crish C, Catania KC (2006) Central visual system of the naked mole-rat (*Heterocephalus glaber*). Anat Rec A Discov Mol Cell Evol Biol 288(2):205–212

Culver DC, Wilkens H (2000) Critical review of the relevant theories of the evolution of subterranean animals. In: Wilkens H, Culver DC, Humphreys WF (eds) Ecosystems of the world: subterranean ecosystems, vol 30. Elsevier, Amsterdam, pp 381–398

Damsgaard C (2016) Acid/base regulation and oxygen transport in the air-breathing fish *Pangasianodon hypophthalmus*. Dissertation, Aarhus University

Davies WL (2011) Adaptive gene loss in vertebrates: photosensitivity as a model case. In: eLS. Wiley, Chichester. doi:10.1002/9780470015902.a002

Deimer P (1977) Der rudimentäre Extremitätengürtel des Pottwals (*Physeter macrocephalus* Linnaeus, 1758), seine Variabilität und Wachstumsallometrie. Z Säugetierkd 42:88–101

Duboué ER, Borowsky RL, Keene AC (2012) β-adrenergic signaling regulates evolutionary derived sleep loss in the Mexican cave fish. Brain Behav Evol 80:233–243

Duboué ER, Keene AC, Borowsky RL (2011) Evolutionary convergence on sleep loss in cavefish populations. Curr Biol 21:671–676

Dufton M, Hall BK, Franz-Odendaal TA (2012) Early lens ablation causes dramatic long-term effects on the shape of bones in the craniofacial skeleton of *Astyanax mexicanus*. PLoS One 7(1–11):e50308

Durand JJ (1971) Recherches sur l'appareil visual du Protèe, *Proteus anguineus* Laurenti, urodèle hypogé. Dissertation, Université Paul Sabatier

Dzwillo M (1982) Curt Kosswig zum Gedächtnis. Mitt Hambg Zool Mus Inst 79:7–17

Eifert C, Farnworth M, Schulz-Mirbach T et al (2014) Brain size variation in extremophile fish: local adaptation versus phenotypic plasticity. J Zool 295(2):143–153

Eigenmann CH (1909) Cave vertebrates of America. Carnegie Inst Wash Publ 104:1–241

Eilertsen M, Drivenes Ö, Edvardsen RB et al (2014) Exorhodopsin and melanopsin systems in the pineal complex and brain at early developmental stages of Atlantic Halibut (*Hippoglossus hippoglossus*). J Comp Neurol 522(18):4003–4022

Eiraku M, Takata N, Ishibashi H et al (2011) Selforganizing optic-cup morphogenesis in three-dimensional culture. Nature 472:51–56

Ekker SC, Ungar AR, Greenstein P et al (1995) Patterning activities of vertebrate hedgehog proteins in the developing eye and brain. Curr Biol 5:944–955

Elipot Y, Hinaux H, Callebert J et al (2013) Evolutionary shift from fighting to foraging in blind cavefish through changes in the serotonin network. Curr Biol 23:1–10

Elliott RE (2015) Cave biodiversity and ecology of the Sierra de El Abra region. In: Keene AC, Yoshizawa M, McGaugh SE (eds) Biology and evolution of the Mexican cavefish. Elsevier, Amsterdam, pp 59–75

Elofsson R, Dahl E (1970) The optic neuropiles and chiasmata of Crustacea. Z Zellforsch 107:343–360

Emerling CA, Springer MS (2014) Eyes underground: regression of visual protein networks in subterranean mammals. Mol Phylogenet Evol 78:260–270

Erckens W, Martin W (1982a) Exogenous and endogenous control of swimming activity in *Astyanax Mexicanus* (Characidae, Pisces) by direct light response and by a circadian oscillator. 1. Analyses of the time-control systems of an epigean river population. Z Naturforsch C 37:1253–1265

Erckens W, Martin W (1982b) Exogenous and endogenous control of swimming activity in *Astyanax Mexicanus* (Characidae, Pisces) by direct light response and by a circadian oscillator. 2. Features of time-controlled behavior of a cave population and their comparison to an epigean ancestral form. Z Naturforsch C 37:1266–1273

Erckens W, Weber F (1976) Rudiments of an ability for time measurement in the cavernicolous fish *Anoptichthys jordani* Hubbs and Innes (Pisces, Characidae). Experientia 32:1297–1299

Ercolini A, Berti R (1975) Light sensitivity experiments and morphology studies. Monit Zool Ital 6:29–43

Espinasa L, Centone DM, Gross JB (2014) A contemporary analysis of a loss-of-function oculocutaneous albinism type II (*Oca2*) allele within the Río Subterráneo *Astyanax* cavefish population. Speleobiol Notes 6:48–54

Espinasa L, Jeffery WR (2006) Conservation of retinal circadian rhythms during cavefish eye degeneration. Evol Dev 8:16–22

Espinasa L, Yamamoto Y, Jeffery WR (2005) Non-optical releasers for aggressive behavior in blind and blinded *Astyanax* (Teleostei, Characidae). Behav Processes 70:144–148

Fack H, Wilkens H (1989) Eye reduction in hybrids and a naturally variable cave form of *Astyanax fasciatus* (Characidae, Pisces). Mém Biospéléol 16:157–161

Falcón J (1999) Cellular circadian clocks in the pineal. Prog Neurobiol 58:121–162

Falcón J, Besseau L, Boeuf G (2007) Molecular and cellular regulation of pineal organ responses. In: Hara T, Zielinski B (eds) Sensory systems neuroscience-fish physiology. Academic/ Elsevier, Amsterdam, pp 243–406. doi:10.1016/S1546-5098(06)25006-4

Falcón J, Migaud H, Muñoz-Cueto JA et al (2010) Current knowledge on the melatonin system in teleost fish. Gen Comp Endocrinol 165:469–482

Foulkes NS, Whitmore D, Vallone D et al (2016) Studying the evolution of the vertebrate circadian clock: the power of fish as comparative models. In: Friedmann T, Dunlap J, Goodwin SF (eds) Genetics, genomics and phenomics of fish, vol 95. Academic, Amsterdam, pp 1–30

Franck D (2012) Curt Kosswig. Ein Forscherleben zwischen Bosporus und Elbe. In: Abhandlungen des Naturwissenschaftlichen Vereins Hamburg, vol 44. Dölling und Galitz, Hamburg, pp 1–216

Freyhof J, Abdullah YS, Ararat K et al (2016) *Eidinemacheilus proudlovei*, a new subterranean loach from Iraqi Kurdistan (Teleostei; Nemacheilidae). Zootaxa 4173(3):225–236

Fricke D (1988) Reaction to alarm substance in cave populations of *Astyanax mexicanus* (Characidae, Pisces). Ethology 76:305–308

Fricke D, Parzefall J (1989) Alarm reaction, aggression and schooling in cave and river populations of *Astyanax fasciatus* and their hybrids. Mém Biospéléol 26:177–182

von Frisch K (1911) Beiträge zur Physiologie der Pigmentzellen in der Fischhaut. Pflügers Arch 138:1–10

von Frisch K (1941) Über einen Schreckstoff der Fischhaut und seine biologische Bedeutung. Z Vgl Physiol 29:46–145. doi:10.1007/BF00304445

Fujii R (2000) The regulation of motile activity in fish chromatophores. Pigment Cell Res 13:$32#300–319

Gregson JNS, Burt de Perera T (2007) Shoaling in eyed and blind morphs of the characin *Astyanax fasciatus* under light and dark conditions. J Fish Biol 70:1615–1619. doi:10.1111/j.1095-8649.2007.01430.x

Gross JB, Borowsky RL, Tabin CJ (2009) A novel role for *Mc1r* in the parallel evolution of depigmentation in independent populations of the cave fish, *Astyanax mexicanus*. PLoS Genet 5:1–14

Gross JB, Krutzler AJ, Carlson BM (2014) Complex craniofacial changes in blind cave-dwelling fish are mediated by genetically symmetric and asymmetric loci. Genetics 196(4):1303–1319
Gross JB, Protas M, Conrad M et al (2008) Synteny and candidate gene prediction using an anchored linkage map of *Astyanax mexicanus*. Proc Natl Acad Sci USA 105:20106–20111
Gross JB, Wilkens H (2013) Albinism in phylogenetically and geographically distinct populations of *Astyanax* cavefish arises through the same loss-of function *Oca2* allele. Heredity 111:$32#122–130
Gross JM, Perkins BD (2008) Zebrafish mutants as models for congenital ocular disorders in humans. Mol Reprod Dev 75:547–555
Hara TJ (1994) The diversity of chemical stimulation in fish olfaction and gustation. Rev Fish Biol Fish 4:1–35
Hausberg C (1995) Das Aggressionsverhalten von *Astyanax fasciatus* (Cuvier 1819; Characidae, Teleostei): Ontogenie, Genetik und Evolution bei der epigäischen und hypogäischen Form. Dissertation, University of Hamburg
Herwig HJ (1976) Comparative ultrastructural investigations of the pineal organ of the blind cave fish *Anoptichthys jordani*, and its ancestor, the eyed river fish, *Astyanax mexicanus*. Cell Tissue Res 167:297–324
Hinaux H, Blin M, Fumey J et al (2015) Lens defects in *Astyanax mexicanus* cavefish: evolution of crystallins and a role for alphaA-crystallin. Dev Neurobiol 75(5):505–521
Hinaux H, Devos L, Blin M et al (2016) Sensory evolution in blind cavefish is driven by early embryonic events during gastrulation and neurulation. Development 143:4521–4532. doi:10.1242/dev.141291
Hinaux H, Pottin K, Chalhoub H et al (2011) A developmental staging table for *Astyanax mexicanus* surface fish and Pachon cavefish. Zebrafish 8:155–165
Hinaux H, Poulain J, Da Silva C et al (2013) De novo sequencing of *Astyanax mexicanus* surface fish and Pachón cavefish transcriptomes reveals enrichment of mutations in cavefish putative eye genes. PLoS One 8(1):e53553. doi:10.1371/journal.pone.0053553
Hoffman S, Hausberg C (1993) The aggressive behavior of the Micos cave population (*Astyanax fasciatus*, Characidae, Teleostei) after selection for functional eyes in comparison to an epigean one. Mém Biospéléol 20:101–103
Hubbs CL, Innes WT (1936) The first known blind fish of the family Characidae: a new genus from Mexico. Occ Pap Mus Zool Univ Mich 342:1–7
Hüppop K (1986a) Oxygen consumption of *Astyanax fasciatus* (Characidae, Pisces): a comparison of epigean and hypogean populations. Environ Biol Fishes 17:299–308
Hüppop K (1986b) Comparative early life history in surface and cave fish (*Astyanax fasciatus*). Bull Zool 35(Suppl):104
Hüppop K (1987) Food-finding ability in cave fish (*Astyanax fasciatus*). Int J Speleol 16 (1–2):$32#59–66
Hüppop K (1989) Genetic analysis of oxygen consumption rate in cave and surface fish of *Astyanax fasciatus* (Characidae, Pisces). Further support for the neutral mutation theory. Mém Biospéléol 16:163–168
Hüppop K (2000) How do cave animals cope with the food scarcity in caves? In: Wilkens H, Culver DC, Humphreys WF (eds) Ecosystems of the world: subterranean ecosystems, vol 30. Elsevier, Amsterdam, pp 159–188
Hüppop K (2012) Adaptation to low food. In: White WB, Culver DC (eds) Encyclopedia of caves, 2nd edn. Elsevier, Amsterdam, pp 1–9
Hüppop K, Wilkens H (1991) Bigger eggs in subterranean *Astyanax fasciatus*. Z Zool Syst Evol 29:280–288
Idda ML, Bertolucci C, Vallone D et al (2012) Circadian clocks: lessons from fish. In: Kalsbeek A, Merrow N, Roenneberg T et al (eds) Progress in brain research, vol 199. Elsevier, Amsterdam, pp 41–57. doi:10.1016/B978-0-444-59427-3.00003-4
Jeffery WR (2001) Cavefish as a model system in evolutionary developmental biology. Dev Biol 231:1–12

Jeffery WR (2005) Adaptive evolution of eye degeneration in the Mexican blind cavefish. J Hered 96:185–196
Jeffery WR, Strickler AG, Yamamoto Y (2003) To see or not to see: evolution of eye degeneration in the Mexican blind cavefish. Integr Comp Biol 43:531–541
Johns PR (1977) Growth of the adult goldfish eye III. Source of the new retinal cells. J Comp Neurol 176:348–358
Kanter MJ, Coombs S (2003) Rheotaxis and prey detection in uniform currents by Lake Michigan mottled sculpin (*Cottus bairdi*). J Exp Biol 206:59–70
Kish PE, Bohnsack BL, Gallina DD (2011) The eye as an organizer of craniofacial development. Genesis 49(4):222–230. doi:10.1002/dvg.20716
Kos M, Bulog B (2000) The ultrastructure of photoreceptor cells in the pineal organ of the blind cave salamander, *Proteus anguinus* (Amphibia, Urodela). Pflügers Arch 439(suppl 1):$32#175–177. doi:10.1007/s004240000136
Kosswig C (1935) Genotypische und phänotypische Geschlechtsbestimmung bei Zahnkarpfen. V. Ein X (Z)-Chromosom als Y-Chromosom in fremdem Erbgut. Roux'Arch Entwickl Mech 133:118–139
Kosswig C (1949) Phänomene der regressiven Evolution im Lichte der Genetik. Commun Fac Sci Univ Ankara 8:110–150
Kosswig C (1967) Über das Tempo evolutorischer Prozesse. Zool Beitr 13:441–450
Kosswig C, Kosswig L (1940) Die Variabilität bei *Asellus aquaticus*, unter besonderer Berücksichtigung der Variabilität in isolierten unter- und oberirdischen Populationen. Rév Fac Sci Istanbul 5:1–55
Kosswig C (1964) Polygenic sex determination. Experientia 20:190–199
Kotrschal K (1992) Quantitative scanning electron microscopy of solitary chemoreceptor cells in cyprinids and other teleosts. Environ Biol Fishes 35:273–282
Kotrschal K (1996) Solitary chemosensory cells: why do primary aquatic vertebrates need another taste system? Trends Ecol Evol 11:110–114
Kotrschal K (2000) Taste(s) and olfaction(s) in fish: a review of specialized subsystems and central integration. Pflügers Arch—Eur J Physiol 439:178–180
Kowalko JE, Rohner N, Linden TA et al (2013a) Convergence in feeding posture occurs through different loci in independently evolved cave populations of *Astyanax mexicanus*. Proc Natl Acad Sci USA 110:16933–16938
Kowalko JE, Rohner N, Rompani SB et al (2013b) Loss of schooling behavior in cavefish through sight-independent and sight-dependent mechanisms. Curr Biol 23:1874–1883
Krishnan J, Rohner N (2017) Cavefish and the basis for eye loss. Philos Trans R Soc B 372:20150487. doi:10.1098/rstb.2015.0487
Kvon EZ, Kamneva OK, Melo US et al (2016) Progressive loss of function in a limb enhancer during snake evolution. Cell 167:633–642
Kwan JW, Lee MJ, Mack AF et al (1996) Nonuniform distribution of cell proliferation in the adult teleost retina. Brain Res 712:40–44
Lamb TD, Collin SP, Pugh EN Jr (2007) Evolution of the vertebrate eye: opsins, photoreceptors, retina and eye cup. Nat Rev Neurosci 8(12):960–976. doi:10.1038/nrn2283
Lande R (1981) The minimum number of genes contributing to quantitative variation between and within populations. Genetics 99:541–553
Langecker TG (1992a) Light sensitivity of cave vertebrates. Behavioral and morphological aspects. In: Camacho AI (ed) The natural history of biospeleology, Monografias del Museo National de Ciencias Naturales. Graficas Mar-Car, Madrid, pp 295–326
Langecker TG (1992b) Persistence of ultrastructurally well developed photoreceptor cells in the pineal organ of a phylogenetically old cave-dwelling population of *Astyanax fasciatus* Cuvier, 1819 (Teleostei, Characidae). Z Zool Syst Evol 30:287–296
Langecker TG (1993) Genetic analysis of the dorsal light reaction in epigean and cave-dwelling *Astynnax fasciatus* (Teleostei, Characidae). Ethol Ecol Evol 3:357–364

Langecker TG (2000) The effects of continuous darkness on cave ecology and cavernicolous evolution. In: Wilkens H, Culver DC, Humphreys WF (eds) Ecosystems of the world: subterranean ecosystems, vol 30. Elsevier, Amsterdam, pp 135–157

Langecker TG, Neumann B, Hausberg C et al (1995) Evolution of the optical releasers for aggressive behavior in cave-dwelling *Astyanax fasciatus* (Teleostei, Characidae). Behav Processes 34:161–168

Langecker TG, Wilkens H (1992) Comparative ultrastructural studies on the pineal organ of the Mexican catfish *Rhamdia laticauda* Heckel, 1858 and one of its cave-dwelling derivates (Pimelodidae, Teleostei). Acta Zool 73(4):247–253

Langecker TG, Schmale H, Wilkens H (1993) Transcription of the opsin gene in degenerate eyes of cave-dwelling *Astyanax fasciatus* (Teleostei, Characidae) and of its conspecific epigean ancestor during early ontogeny. Cell Tissue Res 273:183–192

Langenberg T, Kahana A, Wszalek JA et al (2008) The eye organizes neural crest cell migration. Dev Dyn 237:1645–1652

Lucas RL, Peirson SN, Berson DM et al (2014) Measuring and using light in the melanopsin age. Trends Neurosci 37:1–9. doi:10.1016/j.tins.2013.10.004

Lüling KH (1954) Untersuchungen am Blindfisch *Anoptichthys jordani* Hubbs and Innes (Characidae). 11. Beobachtungen und Experimente an *Anoptichthys jordani* zur Priifung der Einstellung zum Futter, zum Licht und zur Wasserturbulenz. Zool Jahrb Abt Zool Phys 65:9–62

Ma L, Parkhurst A, Jeffery WR (2014) The role of a lens survival pathway including sox2 and αA-crystallin in the evolution of cavefish eye degeneration. Evodevo 5(1):28

Matt N, Dupe V, Garnier JM (2008) Retinoic aciddependent eye morphogenesis is orchestrated by neural crest cells. Development 132:4789–4800

McCauley DW, Hixon E, Jeffery WR (2004) Evolution of pigment cell regression in the cavefish *Astyanax*: a late step in melanogenesis. Evol Dev 6:209–218

McGaugh SE, Gross JB, Aken B et al (2014) The cavefish genome reveals candidate genes for eye loss. Nat Commun 5:1–10

Meng F, Braasch I, Phillips JB et al (2013) Evolution of the eye transcriptome under constant darkness in *Sinocyclocheilus* Cavefish. Mol Biol Evol 30:1527–1543

Missal J (1994) Vergleichende Untersuchungen zur Melatoninsekretion isolierter Pinealorgane epigäischer und hypogäischer Populationen von *Astyanax fasciatus* (Cuvier 1819) (Characidae, Pisces). Dissertation, University of Hamburg

Mitchell RW, Russell WH, Elliott WR (1977) Mexican eyeless characin fishes, genus *Astyanax*: environment, distribution, and evolution. Special publications of the Museum Texas Tech University, 12. Texas Tech Press, Lubbock, pp 1–89

Montgomery JC, Coombs S, Baker CF (2001) The mechanosensory lateral line system of the hypogean form of *Astyanax fasciatus*. Environ Biol Fishes 62:87–96

Montgomery JC, Windsor S, Bassett D (2009) Behavior and physiology of mechanoreception: separating signal and noise. Integr Zool 4:3–12

Moore CM, Roberts RB (2013) Polygenic sex determination. Curr Biol 23:R510–R512

Moran D, Softley R, Warrant EJ (2014) Eyeless Mexican cavefish save energy by eliminating circadian rhythm in metabolism. PLoS One 9(9):e107877. doi:10.1371/journal.pone.0107877

Moran D, Softley R, Warrant EJ (2015) The energetic cost of vision and the evolution of eyeless Mexican cavefish. Sci Adv 1(8):e1500363. doi:10.1126/sciadv.1500363

Nakano T, Ando S, Takata N et al (2012) Self-formation of optic cups and storable stratified neural retina from human ESCs. Cell Stem Cell 10:771–785. doi:10.1016/j.stem.2012.05.009

Negishi K, Stell WK, Takasaki Y (1990) Early histogenesis of the teleostean retina; studies using a novel immunochemical marker, proliferating cell nuclear antigen (PCNA/cyclin). Dev Brain Res 55:121–125

Niemiller ML, Fitzpatrick BM, Shah P et al (2012) Evidence for repeated loss of selective constraint in rhodopsin of amblyopsid cavefishes (Teleostei: Amblyopsidae). Evolution 67:$32#732–748

Niemiller ML, Poulson TL (2010) Studies of the Amblyopsidae: past, present, and future. In: Trajano E, Bichuette ME, Kapoor BG (eds) The biology of subterranean fishes. Science Publishers, New Hampshire, pp 169–280

O'Quin K, Doshi P, Lyon A et al (2015) Complex evolutionary and genetic patterns characterize the loss of scleral ossification in the blind cavefish *Astyanax mexicanus*. PLoS One 10 (12):$32#1–19. e0142208. doi:10.1371/journal.pone.0142208

Ogawa K, Marui T, Caprio J (1997) Bimodal (taste/tactile) fibers innervate the maxillary barbel in the channel catfish. Chem Senses 22(4):477–482

Otteson DC, D'Costa AR, Hitchcock RF (2001) Putative stem cells and the lineage of rod photoreceptors in the mature retina of the goldfish. Dev Biol 232:62–76

Parry JWL, Peirson S, Wilkens H et al (2003) Multiple photopigments from the Mexican blind cavefish, *Astyanax fasciatus*: a microspectrophotometric study. Vision Res 43:31–41

Parzefall J (1970) Morphologische Untersuchungen an einer Höhlenform von *Mollienesia sphenops* (Pisces Poeciliidae). Z Morphol Tiere 68:323–342

Parzefall J (1979) Zur Genetik und biologischen Bedeutung des Aggressionsverhaltens von *Poecilia sphenops* (Pisces, Poeciliidae). Untersuchungen an Bastarden ober- und unterirdisch lebender Populationen. Ethology 50(4):399–422

Parzefall J (1983) Field observations in epigean and cave populations of the Mexican characid *Astyanax mexicanus* (Pisces, Characidae). Mém Biospéléol 10:171–173

Parzefall J (1992) Behavioural aspects in animals living in caves. In: Camacho IA (ed) The natural history of biospeleology. Graficas Mar-Car, Madrid, pp 327–376

Parzefall J (1993) Schooling behaviour in population-hybrids of *Astyanax fasciatus* and *Poecilia mexicana* (Pisces, Characidae and Poeciliidae). In: Schröder JH, Bauer J (eds) Trends in ichthyology: an international perspective. Blackwell Scientific, Oxford, pp 297–303

Parzefall J (2000) Ecological role of aggression in the dark. In: Wilkens H, Culver CD, Humphreys WR (eds) Ecosystems of the world: subterranean ecosystems, vol 30. Elsevier, Amsterdam, pp 221–228

Parzefall J, Fricke D (1991) Alarm reaction and schooling in population hybrids of *Astyanax fasciatus* (Pisces, Characidae). Mém Biospéléol 28:29–32

Parzefall J, Hausberg C (2001) Ontogeny of the aggressive behaviour in epigean and hypogean populations of *Astyanax fasciatus* (Characidae, Teleostei) and their hybrids. Mém Biospéléol 28:153–157

Parzefall J, Trajano E (2010) Behavioral patterns in subterranean fishes. In: Trajano E, Bichuette ME, Kapoor BG (eds) Biology of subterranean fishes. Science, New Hampshire, pp 81–114

Parzefall J, Wilkens H (1972) Artbildung bei Höhlenfischen. Zoomorphology 73:63–79

Patton P, Windsor S, Coombs S (2010) Active wall following by Mexican blind cavefish (*Astyanax mexicanus*). J Comp Physiol A 196:853–867. doi:10.1007/s00359-010-0567-8

Peirson SN, Halford S, Foster RG (2009) The evolution of irradiance detection: melanopsin and the non-visual opsins. Philos Trans R Soc B 364:2849–2865

Peters N, Peters G (1966) Das Auge zweier Höhlenformen von *Astyanax mexicanus* Philippi (Characidae, Pisces). Wilhelm Roux' Arch Entwickl-Mech Org 157:393–414

Peters N, Schacht V, Schmidt W et al (1993) Gehirnproportionen und Ausprägungsgrad der Sinnesorgane von *Astyanax mexicanus* (Pisces, Characinidae). Z Zool Syst Evol 31:144–159

Peters N, Schmidt W, Fricke D (1990) Die Feinstruktur der Kolbenzellen (Schreckstoffzellen) in der Epidermis von *Astyanax mexicanus* Filippi 1853 (Characinidae, Pisces) und seinen Höhlenderivaten "*Anoptichthys*". Int Revue gesamten Hydrobiol 75:257–267

Peters N, Scholl A, Wilkens H (1975) Der Micos-Fisch, Höhlenfisch in statu nascendi oder Bastard? Ein Beitrag zur evolution der Höhlentiere. Z Zool Syst Evolutionsforsch 13:110–124

Pfeiffer W (1960) Über die Schreckreaktion bei Fischen und die Herkunft des Schreckstoffs. Z Vgl Physiol 43:578–614

Pfeiffer W (1967) Die Korrelation von Augengröße und Mittelhirngröße bei Hybriden aus *Astyanax* x *Anoptichthys* (Characidae, Pisces). Roux'Arch Entwickl Mech 150:365–378

Pfeiffer W (1982) Chemical signals in communication. In: Hara T (ed) Chemoreception in fishes. Elsevier, Amsterdam, pp 307–326

Pierce LX, Noche RR, Ponomareva O et al (2008) Novel functions for Period 3 and Exo-rhodopsin in rhythmic transcription and melatonin biosynthesis within the zebrafish pineal organ. Brain Res 1223:11–24

Pitcher TJ (1983) Heuristic definitions of shoaling behavior. Anim Behav 31:611–613

Plath M, Rohde M, Schröder T et al (2006) Female mating preferences in blind cave tetras *Astyanax fasciatus* (Characidae, Teleostei). Behaviour 143:15–32

Popper AN (1970) Auditory capacities of the Mexican blind cave fish *(Astyanax jordani)* and its eyed ancestor *(Astyanax mexicanus)*. Anim Behav 18:552–562

Pottin K, Hinaux H, Retaux S (2011) Restoring eye size in *Astyanax mexicanus* blind cavefish embryos through modulation of the *Shh* and *Fgf8* forebrain organising centres. Development 138:2467–2476

Poulson TL (1963) Cave adaptation in Amblyopsid fishes. Am Midl Nat 41:263–290

Poulson TL, White WB (1969) The cave environment: limestone caves provide unique natural laboratories for studying biological and geological processes. Science 165:971–981

Protas ME, Conrad M, Gross JB et al (2007) Regressive evolution in the Mexican cave tetra, *Astyanax mexicanus*. Curr Biol 17:452–454

Protas ME, Hersey C, Kochanek D et al (2006) Genetic analysis of cave fish reveals molecular convergence in the evolution of albinism. Nat Genet 38:107–111

Protas ME, Patel NP (2008) Evolution of coloration patterns. Annu Rev Cell Dev Biol 24:425–446

Rétaux S, Pottin K, Alunni A (2008) *Shh* and forebrain evolution in the blind cavefish *Astyanax mexicanus*. Biol Cell 100:139–147. doi:10.1042/BC20070084

Riedel G (1997) The forebrain of the blind cave fish *Astyanax hubbsi* (Characidae). Brain Behav Evol 49:20–38

Rodrigues FR (2013) Comparison of brain and cranial nerve morphology between eyed surface fish and blind cave fish of the species *Astyanax mexicanus*. Dissertation, Universidade de Lisboa

Romero A (1985) Ontogenetic change in phototactic reponses of surface and cave populations of *Astyanax fasciatus* (Pisces: Characidae). Copeia 1985:1004–1011

Rummer JL, McKenzie DJ, Innocenti A et al (2013) Root effect hemoglobin may have evolved to enhance general tissue oxygen delivery. Science 340:1327–1329

Sadoglu P (1955) A mendelian gene for albinism in natural cave fish. Experientia 13:394–395

Sadoglu P (1957) Mendelian inheritance in the hybrids between the Mexican blind fishes and their overground ancestor. Verh Dtsch Zool Ges Graz 7:432–439

Sadoglu P, McKee A (1969) A second gene that effects eye and body color in Mexican blind cave fish. J Hered 60:10–14

Schemmel C (1967) Vergleichende Untersuchungen an den Hautsinnesorganen ober- und unterirdisch lebender *Astyanax*-Formen. Z Morphol Tiere 61:255–316

Schemmel C (1977) Zur Morphologie und Funktion der Sinnesorgane von *Typhliasina pearsei* (Hubbs) (Ophidioidea, Teleostei). Zoomorphology 8:191–202

Schemmel C (1974a) Genetische Untersuchungen zur Evolution des Geschmacksapparates bei cavernicolen Fischen. Z Zool Syst Evol 12:169–215

Schemmel C (1974b) Ist die cavernicole Micos-Population von *Astyanax mexicanus* (Characidae, Pisces) hybriden Ursprungs? Mitt Hambg Zool Mus Inst 71:193–201

Schemmel C (1980) Studies on the genetics of feeding behavior in the cave fish *Astyanax mexicanus* f. Anopthichthys. An example of apparent monofactorial inheritance by polygenes. Z Tierpsychol 53:9–22

Schulz-Mirbach T, Ladich F, Riesch R et al (2010) Otolith morphology and hearing abilities in cave- and surface-dwelling ecotypes of the Atlantic molly, *Poecilia mexicana* (Teleostei: Poeciliidae). Hear Res 267(1–2):137–148. doi:10.1016/j.heares.2010.04.001

Shao H, Burrage LC, Sinasac DS et al (2008) Genetic architecture of complex traits: large phenotypic effects and pervasive epistasis. Proc Natl Acad Sci USA 105:19910–19914

Soares D, Niemiller ML (2013) Sensory adaptations of fishes to subterranean environments. BioScience 63:274–283
Stacey N (2003) Hormones, pheromones and reproductive behavior. Fish Physiol Biochem 28:$32#229–235
Stahl BA, Gross JB (2015) Alterations in *Mc1r* gene expression are associated with regressive pigmentation in *Astyanax* cavefish. Dev Genes Evol 225:367–375
Stemmer M, Schuhmacher L-N, Foulkes NS et al (2015) Cavefish eye loss in response to an early block in retinal differentiation progression. Development 142:743–752
Strecker U, Bernatchez L, Wilkens H (2003) Genetic divergence between cave and surface populations of *Astyanax* in Mexico (Characidae, Teleostei). Mol Ecol 12:699–710
Strecker U, Hausdorf B, Wilkens H (2012) Parallel speciation in *Astyanax* cave fish (Teleostei) in Northern Mexico. Mol Phylogenet Evol 62:62–70
Strickler AG, Yamamoto Y, Jeffery WB (2001) Early and late changes in *Pax6* expression accompany eye degeneration during cave fish development. Dev Genes Evol 211:138–144
Strickler AG, Yamamoto Y, Jeffery WB (2007) The lens controls cell survival in the retina: evidence from the blind cavefish *Astyanax*. Dev Biol 311:512–523
Sugimoto M (2002) Morphological color change in fish: Regulation of pigment cell density and morphology. Microsc Res Tech 58:496–503
Sutherland L, Holbrook RI, Burt de Perera T (2009) Sensory system affects orientational strategy in a short-range spatial task in blind and eyed morphs of the fish, *Astyanax fasciatus*. Ethology 115:504–510. doi:10.1111/j.1439-0310.2009.01630.x
Tartellin EE, Frigato E, Bellingham J et al (2012) Encephalic photoreception and phototactic response in the troglobiont Somalian blind cavefish *Phreatichthys andruzzii*. J Exp Biol 215:2898–2903
Teyke T (1990) Morphological differences in neuromasts of the blind cave fish *Astyanax* hubbsi and the sighted river fish *Astyanax mexicanus*. Brain Behav Evol 35:23–30
Varatharasan N, Croll R, Franz-Odendaal T (2009) Taste bud development and patterning in sighted and blind morphs of *Astyanax mexicanus*. Dev Dyn 238:3056–3064
Varjosalo M, Taipale J (2008) Hedgehog: functions and mechanisms. Genes Dev 22:2454–2472
Vinnikow YA (1982) Evolution of receptor cells. Cytological, membraneous and molecular levels. In: Springer GF, Wittmann HG (eds) Molecular biology, biochemistry and biophysics, 34th edn. Springer, Berlin
Weber A (1995) The lateral line system of epigean and cave dwelling catfishes of the genus *Rhamdia* (Pimelodidae, Teleostei) in Mexico. Mém Biospéléol 22:215–225
Weber A (2000) Fish and amphibia. In: Wilkens H, Culver DC, Humphreys WF (eds) Ecosystems of the world: subterranean ecosystems, vol 30. Elsevier, Amsterdam, pp 109–133
Wilkens H (1970a) Beiträge zur Degeneration des Auges bei Cavernicolen. Genzahl und Manifestationsart. Untersuchungen an mexikanischen Höhlenfischen. Z Zool Syst Evol 8 (1):$32#1–47
Wilkens H (1970b) Beiträge zur Degeneration des Melaninpigments bei cavernicolen Sippen des *Astyanax mexicanus* (Filippi). Z Zool Syst Evol 8:173–199
Wilkens H (1970c) Der Bau des Auges cavernicoler Sippen von *Astyanax fasciatus* (Characidae, Pisces). Beitrag zur Problematik degenerativer Evolutionsprozesse. Wilhelm Roux Arch Entwickl Mech Org 166:54–75
Wilkens H (1971) Genetic interpretation of regressive evolutionary processes: studies on hybrid eyes of two *Astyanax* cave populations (Characidae, Pisces). Evolution 25:530–544
Wilkens H (1972) Zur phylogenetischen Rückbildung des Auges Cavernicoler: Untersuchungen an *Anoptichthys jordani (= Astyanax mexicanus)*, Characidae, Pisces. Ann Spéléol 27:$32#411–432
Wilkens H (1976) Genotypic and phenotypic variability in cave animals. Studies on a phylogenetically young cave population of *Astyanax mexicanus* (Fillippi). Ann Spéléol 3:$32#137–148
Wilkens H (1977) Die Rudimente des Rumpfkanals bei kavernicolen Populationen des *Astyanax*. Experientia 33:604

Wilkens H (1982) Regressive evolution and phylogenetic age: the history of colonization of freshwaters of Yucatan by fish and Crustacea. Texas Mem Mus Bull 28:237–243

Wilkens H (1988) Evolution and genetics of epigean and cave *Astyanax fasciatus* (Characidae, Pisces). Support for the neutral mutation theory. In: Hecht MK, Wallace B (eds) Evolutionary biology, vol 23. Plenum, New York, pp 271–367

Wilkens H (2001) Convergent adaptations to cave life in the *Rhamdia laticauda* catfish group (Pimelodidae, Teleostei). Environ Biol Fishes 62:251–261

Wilkens H (2007) Regressive evolution: ontogeny and genetics of cavefish eye rudimentation. Biol J Linn Soc 92:287–296

Wilkens H (2010) Genes, modules and the evolution of cave fish. Heredity 105:413–422

Wilkens H (2016) Genetics and hybridization in surface and cave *Astyanax* (Teleostei): a comparison of regressive and constructive traits. Biol J Linn Soc 118:911–928

Wilkens H, Burns RJ (1972) A new *Anoptichthys* cave population (Characidae, Pisces). Ann Spéléol 27:263–270

Wilkens H, Iliffe TM, Oromí P et al (2009) The corona lava tube, lanzarote: geology, habitat diversity and biogeography. Mar Biodivers 39:155–167. doi:10.1007/s12526-009-0019-2

Wilkens H, Langecker TG, Olcese J (1993) Circadian rhythms of melatonin synthesis in the pineal organ of cave-dwelling *Astyanax fasciatus* (Teleostei: Characidae). Mem Biospeleol 20:279–282

Wilkens H, Meyer M (1992) Eye formation and regression during early ontogeny in cave fish. In: Schroeder JH (ed) Trends in ichthyology. Parey, Hamburg

Wilkens H, Strecker U (2003) Convergent evolution of the cave fish *Astyanax* (Characidae, Teleostei): genetic evidence from reduced eye size and pigmentation. Biol J Linn Soc 80:$32#545–554

Wilkens H, Strecker U, Yager J (1989) Eye reduction and phylogenetic age in ophidiiform cave fish. Z Zool Syst Evol 27:126–134

Windsor SP, Tan D, Montgomery JC (2008) Swimming kinematics and hydrodynamic imaging in the blind Mexican cavefish (*Astyanax fasciatus*). J Exp Biol 211:2950–2959

Yamamoto Y, Byerly MS, Jackman WR et al (2009) Pleiotropic functions of embryonic sonic hedgehog expression link jaw and taste bud amplification with eye loss during cavefish evolution. Dev Biol 330:200–211

Yamamoto Y, Jeffery WR (2000) Central role for the lens in cave fish eye degeneration. Science 289:631–633

Yamamoto Y, Stock DW, Jeffery WR (2003) Development and evolution of craniofacial patterning is mediated by eye-dependent and -independent processes in the cavefish *Astyanax*. Evol Dev 5:435–446

Yamamoto Y, Stock DW, Jeffery WR (2004) Hedgehog signalling controls eye degeneration in blind cavefish. Nature 431:844–847

Yokoyama S, Meany A, Wilkens H et al (1995) Initial mutational steps toward loss of opsin gene function in cavefish. Mol Biol Evol 2:527–532

Yoshizawa M, Goricki S, Soares D et al (2010) Evolution of a behavioral shift mediated by superficial neuromasts helps cavefish find food in darkness. Curr Biol 20:1631–1636

Yoshizawa M, Jeffery WR (2008) Shadow response in the blind cavefish *Astyanax* reveals conservation of a functional pineal eye. J Exp Biol 211:292–299

Yoshizawa M, Jeffery WR, van Netten SM et al (2014) The sensitivity of lateral line receptors and their role in the behavior of Mexican blind cavefish (*Astyanax mexicanus*). J Exp Biol 217:$32#886–895

Yoshizawa M, Robinson BG, Duboué ER (2015) Distinct genetic architecture underlies the emergence of sleep loss and prey-seeking behavior in the Mexican cavefish. BMC Biol 13:$32#15–27

Yoshizawa M, Yamamoto Y, O'Quin KE et al (2012) Evolution of an adaptive behavior and its sensory receptors promotes eye regression in blind cavefish. BMC Biol 10:108–123

Ziv L, Levkovitz S, Toyama R, Falcón J, Gothilf Y (2005) Functional development of the zebrafish pineal gland: lightinduced expression of period 2 is required for onset of the circadian clock. J Neuroendocrinol 17:314–320

# Mechanisms of Regressive Evolution

# 7

**Abstract**

In the face of the neodarwinian paradigm of selection as an agent exclusively claimed to explain evolutionary processes in recent days, almost countless efforts have been made to prove that selection plays an important role in rudimentation processes in cave animals. Efforts mostly focus on the eye and very often pleiotropy is looked upon as being responsible. For example, eye reduction is claimed to antagonistically drive the improvement of taste or the lateral line sense by pleiotropy. However, this could not be confirmed by crossing analysis or by quantitative trait loci mapping. Energy savings have been suggested as another selection factor. This hypothesis implies that all cave species would have to suffer from food limitation. This attempt ignores the fact that the majority of tropical caves, and even some subtropical ones, abound in energy supply. Nonetheless, the traits, having become functionless in the respective cave species occurring in these habitats, regress. Thus, energy limitation is not able to explain regressive evolution of biologically functionless traits in general, or in particular that of the eye in cave species. In fact, independent inheritance of traits suggests that *Astyanax* cave fish are subjected to mosaic evolution.

Besides, selection variability plays a central role in Darwin's concept of evolution. Even in the 1930s, the German evolutionary geneticist Curt Kosswig recognized the importance of the variability exhibited by biologically functionless regressive traits for the interpretation of regressive evolution. According to him, in such traits high phenotypic variability arises and is continuously exhibited for a longer time because a biological function no longer exists. He proposed that this variability was due to the absence of stabilizing selection, because regressive mutations are no longer eliminated and traits become reduced over time merely by their accumulation. Thus, variability and loss are correlated. This so-called "Neutral Mutation Theory" is in accordance with Nei's "Neutral Theory of Molecular Evolution" which applies to molecular evolution. However, Darwin's loss without selection is one of two sides of the same coin, the



H. Wilkens, U. Strecker, *Evolution in the Dark*, DOI 10.1007/978-3-662-54512-6_7

other being "Darwin's gain", in which variability is the basis of constructive evolution. However, usually variability does not exhibit an extent as conspicuous as in the case of functionless cave animal traits. There are only a few examples in which variability, for a relatively short period of time, becomes obvious during evolution. This occurs during the initial phase of processes of adaptive radiation, as can be observed in a series of fish species flocks.

## 7.1 Deleterious Risk

As the most self-evident explanation of eye regression, it was hypothesized that an exposed eye would be a deleterious risk in darkness (Barr 1968; Strickler et al. 2007). This hypothesis is rather improbable, though, because as a rule species colonizing caves are nocturnal. Such so-called troglophile species are already preadapted to a partial life in darkness, are mostly negatively phototactic, and exhibit only small eyes for detecting light. They have already developed improved senses such as taste, olfaction, or orientation, which enable them to orientate and live in complete darkness, too. The best examples of this are provided by catfish (Wilkens 2001). In surface *Astyanax*, a comparatively well-developed lateral line sense facilitated orientation in darkness (see Sect. 6.5).

## 7.2 Food Limitation and Energy Economy

Energy economy is often looked upon as a driving force of regressive evolutionary processes, because in general caves are considered food poor and energy limited (Hüppop 2000, 2012). However, this does not generally correspond to reality and is misleading. Tropical and some subtropical caves may abound with bat or bird guano and flood detritus, providing rich, stable food resources, and are inhabited by large numbers of eye- and pigment-reduced species (Deharveng and Bedos 2000; Fernandes et al. 2016; Gnaspini and Trajano 2000). For example, such conditions were described for a food-rich sulphidic Mexican cave, in which the live-bearing toothcarp *Poecilia mexicana* exhibits highly variable eye sizes and all degrees of pigmentation (Fig. 3.14) (Langecker et al. 1996; Palacios et al. 2016; Peters and Peters 1973, 1968). Direct evidence for the uncoupling of energy input and regressive evolutionary processes is provided by cave-living plant hoppers. Although they suck on the roots of trees growing outside caves, and thus are directly dependent on primary production for their energy, they exhibit reduced eyes and dark pigmentation (Hoch 2000). Whereas a lack of light is an exclusive characteristic of caves, a small energy supply is not. Therefore, food limitation is a potential secondary independent environmental factor influencing the evolution of cave animals.

In particular, eye reduction was the focus of many studies and used as the prime example when trying to prove the role of energy economy as a driving selective force. It has never been considered that several behavioural traits such as schooling, visually released aggressive behaviour, or the dorsal light reaction are no longer being performed in darkness by the surface ancestor because of the missing releaser, but nonetheless are reduced in the cave populations (see Sects. 6.12, 6.13, and 6.18). As concerns the eye, several authors claim that the high energetic costs of neural cell metabolism could be a strong selective force (Borowsky and Cohen 2013; Moran et al. 2015; Niven et al. 2007; Protas et al. 2007). However, all these studies were performed with light-bred specimens, which does not compare to a cave environment. Comparison of eyes of fish living in permanent darkness and those kept under normal daylight conditions revealed different degrees of retinal histological differentiation. For example, in the phylogenetically young cave fish *P. mexicana* (Poeciliidae), which still exhibits externally visible eyes with variable size, specimens collected in the cave showed considerable malformations of the retina, whereas their offspring hatched and kept at 12-h light/12-h dark conditions developed completely intact and functional retinas (Parzefall 2001; Peters and Peters 1968; Peters et al. 1975). The same was revealed for the cave salamander *Typhlotriton spelaeus* (Besharse and Brandon 1974). Also, adult *Astyanax* surface specimens bred and kept in continuous darkness in a laboratory exhibit retinas that are about one third thinner than those reared in light, and the number of visual cell nuclei building the outer nuclear layer is reduced by 20% (Wilkens 1988). This is corroborated by the differing eye sizes of wild Micos cave fish and their laboratory offspring (Wilkens 1976). Thus, experiments performed with fish bred in light do not consider that in caves their entire ontogenetic development takes place in complete darkness, without any light stimulus. It is known that light stimulus is important for the formation of the visual apparatus (Zeutzius et al. 1984; Yang et al. 1988a, b). Such results are not comparable with those from fish existing under the exclusion of light. Most importantly, the phenotypic variability of eye size in cave species brings into question the lack of directional selection in eye loss (see Sect. 7.8). The combination of all these facts challenges whether the energy gain of eye reduction is important.

## 7.3 Quantitative Trait Loci Polarity Test for Selection (Orr's Sign Test)

Applying the quantitative trait loci (QTL) test developed by Orr (1998), it was claimed that cave alleles at every eye QTL caused size reduction and therefore argued that they were submitted to directional selection (Borowsky 2015; Krishnan and Rohner 2017; Protas et al. 2007). In contrast, QTL polarities for melanophores were mixed and thus consistent with neutral evolution. In fact, Orr's sign test itself is unable to prove selection (be it positive selection, relaxed selection, or whatever) by looking at QTL allele patterns between populations. Russell Lande states:

As I understand the model and assumptions of this method, it only deals with the divergent aspect of selection between two populations descended from a common ancestor, and is ambiguous about selection within them. In this case it is obvious that selection maintains functional eyes in surface-dwelling fish, and it is safe to assume that most mutations in functional eyes, especially mutations of substantial effect that are likely to be detected in QTL experiments, will tend to reduce them (consistent with the empirical conclusion of Muller 1950) and are deleterious in surface populations. On colonizing a cave environment, even if selection there is merely absent and not actually against eyes, the mutational bias toward eye reduction (in both originally existing and new mutations), by itself or particularly when combined with genetic drift, will tend to fix eye reduction alleles. With a sufficient number of QTL the sign test will conclude that divergent selection has caused the difference. This is true in the sense that selection favors relatively large eyes in surface populations in comparison to cave populations. One can conclude that a significant result in Orr's sign test says nothing about whether within cave populations selection favors reduced eyes or is absent (Lande, personal communication).

## 7.4 Pleiotropy

Pleiotropy in its classical sense is defined as a genetic action that occurs where one gene is responsible for two or more unrelated traits. The classical example is provided by the adaptive advantage of individuals heterozygous for the sickle cell disease with respect to malaria. Recently theories of eye regression, as originally suggested by Barr (1968), have come to the fore again. According to this theory, regressive evolutionary processes were driven by the same genes, which improve cave adaptive traits such as taste.

### 7.4.1 Pleiotropy of Sonic Hedgehog (*Shh*) Genes

In *Astyanax* cave fish embryos, eye primordia were found to degenerate under the influence of hyperactive *sonic hedgehog* (*shh*) signalling (Yamamoto et al. 2004). As *shh* also plays an important role in oral jaw and taste bud development, it was suggested that *shh* signalling might increase traits like oral jaw size and taste organ number in cave fish and that by antagonistic pleiotropy could be responsible for eye loss (Jeffery 2005; Yamamoto et al. 2009). Attempts were made to prove this by studying F3-crossing hybrids, the third generation between surface and cave fish. F3-hybrid specimens were found in which small eyes were combined with many taste buds and vice versa. However, F3 crossings are not an appropriate tool to detect pleiotropy, because such crossings are the offspring of F2 specimens, the genetic background of which is undefined and unknown. F2 specimens only represent a selected part of the whole genetic variability. The scientific standard for proving pleiotropy would have been in an F2 crossing, which derives from F1 hybrids that possess a complete haplome from each parental form. This hypothesis is finally rejected by QTL analysis, which revealed that none of the multiple QTL underlying eye regression is located near a known *shh* locus (Protas et al. 2007).

In a similar pleiotropic approach, attempts were made to show that, due to hyperactive *shh* signalling, not only would the cave fish eye anlage become smaller, but the forebrain including the hypothalamus, where crucial neural centres controlling aggressive behaviour are located, would increase in size (Menuet et al. 2007). A link was claimed to exist between a modified specification of the hypothalamus and the differential control of aggressive behaviour, which was assumed to be missing in the cave fish (Rétaux et al. 2008). This interpretation does not take into account that the cave fish still exhibit aggressive interaction (see Sect. 6.12). Furthermore, the existence of an increase of forebrain size is controversial. Whereas Peters et al. (1993) found that its superior area size was enlarged, calculation of the total volumes of the forebrain and the hypothalamus did not state an enlargement in cave fish compared with the surface fish (Rodriguez 2013) (see Sect. 6.22).

Above all, these conclusions do not sufficiently consider the differential gene expression and the cell-fate-determining activity of *shh* as a morphogen, which depends on its concentration and is time dependent. Thus, *shh* is definitely involved in trait morphogenesis, but in diverging circumstances. Additionally, F2 crossings between surface and cave fish revealed that the above-mentioned traits inherit independently (see Sects. 6.21.5, 6.21.6, 7.5) (Wilkens 2010). Thus, the alleged pleiotropic links between regressive eye and constructive traits like oral and taste bud number, or hypothalamus size and aggressive behaviour, are questionable (Wilkens 1988, 2010, 2016).

### 7.4.2 Pleiotropy of Neuromodulation (Melanin-Catecholamine Trade-Off Hypothesis)

Neuromodulation has received growing interest in being indirectly responsible for cave regressive and constructive processes of behavioural traits (Katz and Lillvis 2014). Neuromodulators like dopamine, serotonin, acetylcholine or histamine secreted by a small group of neurons regulate diverse populations of neurons. The neuromodulators diffuse through large areas of the nervous system affecting multiple neurons. The broad neuromodulatory control exerted by monoamines (MOA) in vertebrate brains explains why they have been implicated in a variety of functions and processes: attention and arousal (noradrenaline, NA), defence and anxiety (dopamine, DA; serotonin, 5-HT), stress and emotions (DA, NA, 5-HT), reward (DA), mood (5-HT), appetite and energy expenditure, sleep/wakefulness (5-HT, NA, melatonin), and motor control (DA, 5-HT).

Neuromodulatory systems provide sites at which natural selection can act to alter the output of neural circuits and thereby play a role in the evolution of behaviour (Katz 2011; Katz and Lillvis 2014). By changes in receptor expression levels or release of neuromodulatory substances affecting the expression of genes via mutations in regulatory regions, a mechanism is provided to alter neuromodulatory signalling. It is suggested that parallel evolution of neuromodulatory signalling systems is responsible for repeated evolution of specific behaviours of different

*Astyanax* cave fish populations (Hinaux et al. 2015). For example, it is proposed that parallel evolution of certain behavioural traits like feeding posture in different *Astyanax* cave fish populations causes slightly different angles of feeding posture between the strongly eye-and pigment-reduced (SEP) and the variably eye-and pigment-reduced (VEP) cave fish (Katz and Lillvis 2014).

Neuromodulation was thought to be responsible for changes in a whole range of different behaviours (Elipot et al. 2013). In cave fish brains, high levels of content and neurotransmission indexes for serotonin, dopamine, and noradrenaline, but low MAO activity was detected. By analyzing the MAO coding sequence a mutation was identified, which is assumed to be responsible for this. The same mutated allele was found in the SEP Pachón, Sabinos, and Tinaja cave fish. Low MAO activity was claimed to be advantageous for cave life, although in humans such condition would be pathological. It was proposed to provide the genetic basis for several aspects of the so-called "*Astyanax* cave fish behavioural syndrome", which includes loss of aggression, schooling, and sleep, and increased food searching. Curiously, the MAO mutation was not found in the VEP Micos and Molino cave fish, although they exhibit similar behavioural traits.

In another approach, it was found that L-tyrosine, dopamine, and norepinephrine levels were enhanced in larval pre-feeding stages and adult brains of the Pachón cave fish (Bilandzija et al. 2013; Jeffery et al. 2015). This was attributed to the *oca2* loss-of-function albino mutation, which therefore was suggested to be an evolutionary benefit, because the enhancement of catecholamine levels was made possible by surplus L-tyrosine deriving from the block in melanin synthesis as its precursor. It would be a possible target of natural selection due to the importance of these compounds as neurotransmitters and paracrine factors. The inverse relationship between the block in melanin synthesis and the enhancement in catecholamine synthesis is hypothesized to be an example of "secondary pleiotropy", a type of pleiotropy in which a single mutated gene in a defined biochemical pathway has multiple phenotypic consequences. The resultant increases in serotonin and catecholamine signalling were hypothesized to play a role in modulating behaviours like reduced schooling, aggression, or sleep that could "render cavefish more successful in their natural environment" (Bilandzija et al. 2013).

Both hypotheses described before assume that the cave fish have elevated brain levels of serotonin and dopamine (and secondarily noradrenaline) indirectly caused by two independent mutations in MAO and *oca2*. However, potential gains of low MAO activity or surplus L-tyrosine are hampered by the fact that traits like schooling and aggressive behaviour are visually released and therefore not performed in darkness by the surface fish. Thus, after the surface fish started to evolve to a cave fish, selection cannot have acted on a "hidden" trait and been responsible for the spreading of these mutations in the various cave fish gene pools. Furthermore, the traits involved in the "behavioural syndrome" like schooling, feeding posture, or sleep have a polygenic basis and do not inherit in a monogenic way, as must be expected when suggesting that their change relies on just one mutation in *oca2* or MAO, respectively. In addition the albino gene inherits independently from those traits "rendering the Pachón cave fish more successful in the cave environment". The hypothetic role of the albino gene is all the more probable given the existence of only three albinotic *Astyanax* cave fish populations among 27 without the albino gene, all of which nonetheless show the same cave behavioural traits mentioned before (Gross et al. 2009; Mitchell et al. 1977; Wilkens 1988, 2010; Wilkens and Strecker 2003).

In summary, neuromodulatory systems provide sites at which natural selection can act and it can be assumed that behavioural traits of *Astyanax* cave fish like feeding posture and sleep have constructively been adapted to cave life under the influence of selection. It is important, however, to differentiate between constructive and regressive traits. In contrast to constructive evolutionary processes, in regressive ones degenerative mutations will strike both the genetic neuromodulatory basis and the one responsible for the performance of the pattern of a certain behavioural trait. It was shown by crossing experiments that regressive mutations have accumulated in visually released traits like schooling and aggressive behaviour or dorsal light reaction (see Sects. 6.18, 6.12, and 6.13). These traits cannot be performed by the surface fish in the dark, because they depend on vision, nor are they performed by the cave fish, because the polygenic basis is deteriorated (Kowalko et al. 2013b; Langecker 1993; Parzefall 1993; Wilkens 1988). It seems much more parsimonious to explain the reduction of behavioural traits not being performed in darkness by the accumulation of regressive mutations due to loss of stabilizing selection than by a single mutation in MAO or *oca2,* respectively, simultaneously influencing neuromodulation of a large number of behaviours. Thus, it is questionable whether melanin-reducing genes like *oca2* (Bilandzija et al. 2013) or the MAO mutation (Elipot et al. 2013) are responsible for the regression of some specific behavioural traits in cave fish.

### 7.4.3 Pleiotropy of Vibration Attraction Behaviour (VAB) and Superficial Neuromasts in the Cave Fish Orbit

Using quantitative genetic QTL analysis, it was hypothesized that the adaptive vibration attraction behaviour (VAB) and a small group of the sensory receptors, superficial neuromasts, occurring within a small area of the cave fish eye orbit (EO) (Figs. 6.2 and 6.3), were genetically correlated with reduced eye size. From this was concluded that natural selection for the enhancement of VAB and EO neuromasts would indirectly promote eye regression in the Pachón cave fish population through an antagonistic relationship involving genetic linkage or pleiotropy among the genetic factors underlying these traits (Yoshizawa et al. 2012). However, this genetic analysis ignores the numerous other QTL that map to, and potentially interact, in the same chromosomal regions and thus fails to establish pleiotropy (Borowsky and Cohen 2013). Furthermore, it was found that neither eye regression itself nor experimental *shh* overexpression induces EO neuromast formation (Yoshizawa et al. 2015). Thus, as VAB is not exhibited in a series of SEP cave populations, which nonetheless carry strongly reduced eyes, a selective advantage of eye loss via improved VAB and increased number of neuromasts within a restricted area of the eye orbit as hypothesized by Yoshizawa et al. (2012) is unproven. This is all the more so since all free neuromasts developed on the cave fish head are inherited as one single unit and obviously rely on the same genetic basis. This is corroborated by the correlation between the numbers of superficial neuromasts on the cave fish head and those in a remote separate position along the maxillary canals underneath the jaw (Fig. 6.2b) (Wilkens 2010).

## 7.5 Independent Inheritance

The crossing analyses of F2 hybrids between surface and the different cave fish populations provide an excellent tool to analyze whether the most diverse constructive or regressive traits exhibit independent inheritance (Fig. 7.1). These results as well as those of QTL studies did not reveal phenotypic correlations between the size of the eye and cave adaptive traits such as the number of taste buds (Protas et al. 2007; Schemmel 1967), amount of fat storage (Hüppop, 1989), feeding posture (Kowalko et al. 2013b; Schemmel 1980), aggressive behaviour (Hofmann and Hausberg 1993), schooling (Kowalko et al. 2013a), or the number of teeth (Protas et al. 2007). There was also no correlation between the eye and jaw size, naris size, or mouth width, which are each larger in the cave fish (Wilkens 2010). All traits studied, including the diverse colour mutations, are inherited independently. Even within single traits like eye or melanophore system, the two associated units (lens and retina or number of melanophores and morphological colour change, respectively) that they consist of are inherited independently (see Sect. 6.23) (Wilkens 1988, 2010, 2016). The hypothesis that increased *shh*-signalling reduces eye size and synchronously amplifies taste bud number and jaw size does not consider the subtle equilibrium that exists in space and time between the different signalling systems orchestrating the development of the different traits (Hinaux et al. 2016). Manipulation of early signalling systems usually results in abnormal phenotypes. Thus, it is doubtful that morphological evolution is due to modifications at this early ontogenetic level.

The observations shown above fit into the concept of mosaic evolution, namely that every species is a mosaic of traits that persist unchanged, whereas some traits change when submitted to selection pressure, exhibiting different intensities or even being relaxed and missing (Wilkens 2010). Mosaic evolution was also found for overall brain structure in mammals and the central visual system of the naked mole rat (*Heterocephalus glaber*), which has selectively lost structures that mediate form vision while those needed for the entrainment of the circadian rhythm are retained (Barton and Harvey 2000; Crish et al. 2006). In African cichlids, there is a close relationship between the relative sizes of various brain structures and the utilization of habitat and prey (Huber et al. 1997). Deep sea fish provide evidence for a great variety of diverse specializations of lens and retina in adaptation to light-poor environments (Ellis 1996). For example, the fish *Ipnops murrayi* (Ipnopidae) has modified its eyes with the retina covering most of the upper surface of the head in order to perceive faint bioluminescent light signals, but a lens no longer exists.

**Fig. 7.1** F2 hybrid specimen between SEP Piedras cave and surface fish exhibiting a mixture of surface and cave traits like loss of guanine in the scales, low melanophore density but no reduced melanin content, and intermediate eye size. *SEP* strongly eye- and pigment-reduced

Other deep sea fishes like the telescope fish *Gigantura chuni* (Giganturidae) possess heads dominated by large, forward-pointing, telescoping eyes with voluminous lenses. Disadvantages would arise from linkages if, for example, the lens could not be enlarged without consequences for the retina. This would impede the evolutionary process of optimal adaptation. From an evolutionary point of view, independent inheritance of traits is suggested to be mostly advantageous.

## 7.6 Evolutionary Rates of Regressive and Constructive Traits

Recent cave-colonizing species like the VEP Micos cave fish offer excellent opportunity to study the evolutionary rates of cave-specific traits. In contrast to the SEP cave fish populations, in which the process of cave evolution may have come to an end an unknown time ago, the VEP cave fish have not yet finished. It was revealed for the Micos cave fish that mean and variability of some of the constructive traits like egg yolk content, fat storage ability, or nose pit size have already achieved the degree characteristic of SEP cave fish (Aspiras et al. 2015; Bibliowicz et al. 2013; Hüppop and Wilkens 1991; Wilkens 1988). However, other constructive traits like number of taste organs and a newly evolving aggressive behaviour elicited by the lateral line sense have not reached a state comparable to that of the SEP cave fish, but are intermediate between surface and SEP cave fish. Regressive traits like eyes, dorsal light reaction, or the visually triggered aggressive behaviour exhibited are also at an intermediate stage. The number of melanophores, as another regressive trait, phenotypically equals that of the surface fish. Nonetheless, its genetic basis has already been submitted to regressive mutations. However, this defect does not yet manifest due to epistatic gene effect (see Sect. 6.23.3). In summary, it is suggested that the evolution of the specific traits in cave species, independently from being regressive or constructive, proceeds at different rates. It is therefore questionable to conclude from the degree of the reduction of a trait like the eye in comparison with that of another one (melanophore pigmentation) whether they are reduced with or without selection as claimed for balitorid cave loaches (Borowsky 2015). Such calculations do not take into consideration that discontinuous epistatic gene effects might falsify the time estimates based on trait manifestation. This calculation is all the more questionable as long as nothing is known about the genetic expression pattern of the two traitsstudied in the balitorid cave loaches. For example, eye size is reduced by a threshold-like 25% amount in the Micos cave fish at the beginning of reduction, from which strong selection pressure and rapid evolution might be concluded. However, further reduction proceeds in many smaller steps, implying a much slower advance of regression (see Sect. 6.23). The evolution of constructive and regressive traits can best be explained by mosaic evolution.

## 7.7 Reversibility of Regressive Evolution

Dollo's law of irreversibility states that evolution is not reversible (Dollo 1893). However, eyes and pigmentation of cave animals have often been claimed as providing examples to rebut this hypothesis (e.g. Dillman et al. 2010). Such a misleading assumption may have arisen from the observation that several pale cave species, when exposed to daylight, become completely black or at least darker. The amphibian *Proteus anguinus* from the Dinarian karst offers one of the most spectacular examples for this. It was also found in the cave fishes *Rhamdia zongolicensis* and *R. reddelli* (Fig. 2.7) (Wilkens 2001) as well as in the Yucatan cave swamp eel *Ophisternon infernale* (Fig. 3.18). In *Astyanax*, phylogenetically younger VEP cave populations like Micos deviate from the phylogenetically older SEP ones by the observation that at light they may get darker again. Also, cave crayfish like *Procambarus reddelli oaxacae* become brownish when kept at daylight. The phenomenon of darkening can be explained by the production of melanin and ommochrome pigments in part still being undisturbed by mutations in these cave species. It is just downregulated in continuous darkness. Therefore, the more recent cave-invading, phylogenetically young populations are still able to synthesize melanin. Independently they can still increase the number of melanophores by morphological colour change when kept at daylight. This by no means counts as an example of reversible evolution. In contrast, however, the majority of cave species, which are represented by a phylogenetically older age, can no longer produce dark pigments and higher numbers of melanophores under such conditions, because the ability has been genetically lost. They have become irreversibly colourless.

Size and histological differentiation of the eye rudiments of phylogenetically old SEP *Astyanax* cave fish from the cave habitat are not different from their offspring bred and kept under daylight. However, the eyes of phylogenetically younger *Astyanax* cave populations like the VEP Micos fish are slightly better developed when hatched at daylight. This was explained by light environmental influence on the still existing visual cell outer segments (Peters et al. 1975; Peters and Peters 1968). It had only minor effect on eye size as a whole, though. However, the "re-appearance" of truly large eyes in cave animals was presented for the amphipod *Gammarus minus* (Culver et al. 1995) and the phylogenetically young VEP *Astyanax* Caballo Moro cave fish (see Sect. 5.3.3) (Espinasa and Borowsky 2000). Both species occur below a lighted karst window, by which cave animals get access to daylight.

In order to explain the occurrence of these larger eyes, it was hypothesized that the eyed *Astyanax* cave fish below the karst window in the Caballo Moro Cave were direct descendants of the blind cave fish population inhabiting the dark parts of this cave that had secondarily genetically "re-acquired" eyes. However, the more probable solution for these observations is provided by the results of crossing analyses between *Astyanax* surface and cave fish. They showed that due to an epistatic genetic threshold, the eye size is increased by a discontinuous step after a minimal number of the remaining not mutated eye genes are recombined (Wilkens 1988, 2010, see Sect. 6.23). In nature, a similar phenomenon has occurred in the variable phylogenetically young VEP Micos cave fish. However, whereas in the dark Micos cave such specimens with larger eyes do not have any selective advantage and are rare, this is different in the lighted karst window of the Caballo Moro cave. Here the eyed specimens are favoured by the lighter condition and concentrate to a population, which can be separated from the rest of the stronger eye-reduced cave population by molecular analysis (Espinasa and Borowsky 2000; Strecker et al. 2012). A comparable explanation is provided for the "re-appearance" of the large eyed *G. minus* amphipodes, which obviously exhibit a similar genetic pattern of eye manifestation (Culver 1987) as the fish *Astyanax*.

In summary, the "re-appearance" of both eyes and melanin pigmentation is exclusively observed in phylogenetically younger cave populations, in which it is characteristic that the genetic regression of these traits has not entirely been completed. Thus, this phenomenon is a consequence of the manifestation of still existing intact eye or melanophore/melanin pigment alleles and not restoration of degenerate ones by "back-mutations". In the case of the eye, the re-appearance is much more improbable because it would require back mutations of a large number of genes. It is furthermore generally not considered that cave species are not exclusively characterized by the reduction of eyes and melanin pigmentation, but that many more traits have been altered by regressive or constructive evolution. Thus, the reversibility to the former ancestor, as claimed by Dollo's law, would also require the "restoration" of many more traits and become even more unrealistic.

## 7.8 Variability and Loss: Neutral Mutation Theory

Besides selection variability plays a central role in Darwin's concept of evolution. It was the idea of the evolutionary geneticist Curt Kosswig (1903-1982) to study the genetic basis of regressive evolution by investigating cave animals. He reiterated that Darwin's ideas were in concert with the current well known facts of genetics: "*The majority of mutations is responsible for damages in phenotype, in particular in case a trait is no longer controlled by selection. Within the bounds of constraints provided by the developmental and genetic backgrounds of an organism regressive mutations are accumulating step by step, by which the ultimate result mentioned by Darwin will finally originate. It is the loss of purifying selection, which is looked upon as being the cause of rudimentation, because deteriorating mutations are no longer eliminated now*" (Kosswig 1935, 1949, 1960a, b, 1976; Kosswig and Kosswig 1940).

Like Kosswig, Nobel Laureate H.J. Muller (1949) embedded the ideas of Charles Darwin into the knowledge of modern genetics: "*But even though organisms may have reached a virtual stopping point in evolution, the mere maintenance of their present structure requires, as we realize today, the persistent operation of natural selection. Darwin, to be sure, stated that a relaxation of selection leads to an increased accumulation of individual variants. But in his time the knowledge was apparently lacking that the relaxation of selection with regard to any character would lead to decay down to the level at which selection does operate, and that an actual cessation of selection for a character would in time lead to its complete disappearance. In fact, the entire organization would deteriorate similarly, if selection in all directions were relaxed. Before the advent of modern mutation study, it was not known that genetic changes in the down-hill direction are in general far more frequent than those which increase or intensify an organ or character. True, Darwin was so astute as to point out that such a principle, if true, would be of great service in explaining the reason for the decline, and more especially for the total disappearance of features which had lost their usefulness, like the eyes and the pigmentation of cave animals*".

The Neutral Mutation Theory (Kosswig 1960a; Wilkens 1988, 2010) incorporates the forward-looking principles and ideas of Darwin with the insights of modern genetics showing that constructive and regressive evolution are not contradictive but basically rely on identical mechanisms: mutation-driven phenotypic variability combined with absent or present selection. In case of the absence of selection, the variability is not eliminated and traits get reduced over time merely by the accumulation of regressive mutations. In contrast, in the presence of selection, mutations negatively influencing a specific trait are eliminated and no variability occurs. This genetic interpretation refines Darwin's explanation of rudimentation, when he proposed that "If, for instance, it could be proven that every part of the organisation tends to vary in a greater degree towards diminution than towards augmentation of size then we should be able to understand how an organ which has become useless would be rendered, independently of the effects of disuse, rudimentary and would at last be wholly suppressed; for the variations towards diminished size would no longer be checked by natural selection". A rising number of studies are corroborating Darwin's assumptions. For example, there has been repeated loss of functional constraint of rhodopsin in amblyopsid cave fishes, as at least three cave lineages have independently accumulated unique loss-of-function mutations. Although several cave lineages still possess functional rhodopsin, they exhibit increased rates of nonsynonymous mutations that have greater effect on the structure and function of rhodopsin compared with those in surface lineages (Niemiller et al. 2012). Kowalko et al. (2013a, b) argue that once vision was impaired by the lack of light, schooling was no longer under selection, and alterations in genes affecting this behaviour would be neutral in consequence. Transcriptome sequencing of embryonic and larval stages of the *Astyanax* surface fish and the Pachón cave fish show a high number of mutations in cave fish putative eye genes may be explained by relaxed selection for vision during evolution in the absence of light (Hinaux et al. 2013).

The Neutral Mutation Theory relies on principles similar to Kimura's "Neutral Theory of Molecular Evolution". This latter theory holds that at the molecular level most evolutionary changes and most of the variation within and between species is not caused by natural selection but by random drift of mutant alleles that are neutral and do not affect the ability of an organism to survive and reproduce (Freese and Yoshida 1965; Kimura 1968; King and Jukes 1969; Nei et al. 2010; Nei 2013). The phenotypic variability of regressive traits, for which cave animals in particular provide many examples, would be one of the rare cases in which random neutral mutations can phenotypically manifest without being eliminated by natural selection normally acting to preserve the functional capability of a feature.

### 7.8.1 Variability of Regressive Traits

A generally observable and outstanding feature of functionless regressive traits is their long-lasting genetic and phenotypic variability in size and form. It occurs in all taxonomic orders and is shown, for example, by the reduced hind wings of flightless carabid ground beetles (Fig. 7.2) as well as by the rudimentary hip bones of whales (Fig. 2.2). However, it is particularly obvious in cave species. The important specific characteristic of variability is generally not adequately considered. One of the first to call special attention to it was Eigenmann (1909), who observed it while studying the eyes of the Cuban bythidid cave cusk eels, genus *Lucifuga* (Fig. 6.38). Kosswig and Kosswig (1940) recognized its general relevance for the interpretation of regressive evolutionary processes and were the first to expound the problems of the variability of functionless traits and to focus on the implication for evolution (Culver et al. 1995). They started studying the variable eyes and pigmentation in cave populations of the cave wood louse *Asellus kosswigi* (Verovnik et al. 2009; Konec et al. 2016), performing morphological as well as crossing experiments for genetic analysis (Kosswig and Kosswig 1940). The genetic basis of the variability of eyes and pigmentation of this cave isopod was later stated by QTL analyses (Protas et al. 2011).

Further examples of variability are provided by cave species of all taxonomic orders. For example, the eye rudiments of the diverse Caribbean cave shrimps genus *Typhlatya* (Fig. 3.12) (Juberthie-Jupeau 1976) and those of the cave mysid *Heteromysoides cotti* (Meyer-Rochow and Juberthie-Jupeau 1987) as well as those of the cave catfish genus *Rhamdia* (Wilkens 2001), the amblyopsid cave fish (Poulson and White 1969), and the Bahamian bythidid cave cusk eels are variable in structure and size (Wilkens et al. 1989) (Figs. 3.17 and 6.38). The live-bearing cave-living tooth carp *Poecilia mexicana* provides another interesting example (Peters and Peters 1973). Its eyes exhibit a gradient of decreasing size and increasing variability within the same cave from outside, where the ancestral well-eyed surface fish lives, into deeper cave regions (Fig. 3.14).

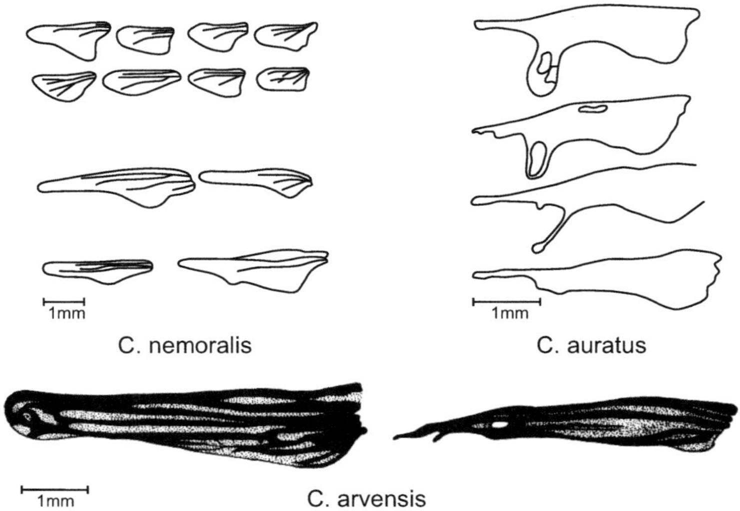


**Fig. 7.2** Variability of the reduced hind wings of different flightless *Carabus* ground beetle species (Carabidae, Coleoptera) in size, structure, and form (adapted from Wilkens et al. 1979)

In the *Astyanax* cave fish, variability of regressive features is manifold, too. It exists within populations as well as between them (Fig. 6.39). Eye variability is particularly obvious in the phylogenetically young VEP cave fish populations like Micos, which exhibit externally visible eyes, the sizes of which encompass almost half of those developed in F2 crossings between surface fish and the phylogenetically old SEP cave fish (Figs. 5.4, 5.7 and 6.55a). In all cave fish, individual left-right asymmetry of size and histological differentiation of the eyes occurs (Fig. 6.48). However, variability of regressive features is not restricted to morphological traits like eyes and pigmentation but may also develop in behavioural ones. For example, the rudimentary visually released aggressive behaviour as well as the dorsal light reaction is submitted to it in the phylogenetically young VEP Micos population (Langecker et al. 1995, see Sects. 6.12 and 6.13).

Unlike in cave species, rudimentation mostly starts with an evolutionary process of change during which the function of a specific trait is gradually substituted by another one. During this phase, phenotypic variability will not appear, because stabilizing selection is still in play. Only at the end of this development, when the original function of a trait has completely been transferred to the new one, may variability arise for the one being reduced. For example, during phylogeny the hind legs of the ancestral whales were for a long time still functionally involved in promotion and could only completely regress after the tail fin had taken over that function. Only at that moment genotypic and phenotypic variability arose (Fig. 2.2). In contrast, it is an extraordinary characteristic that when species become permanently cave living, traits that depend on light to function abruptly lose their biological function without substitution.

In accordance with the recapitulation theory, which claims that during ontogeny phases of phylogeny are recapitulated, this may also be observed during the ontogeny of regressive traits; for example, variability does not exist during early ontogeny of the cave fish eye, but arises later. This could be exemplified in cave *Astyanax*. Whereas at early stages eye variability of the cave fish does not surpass that of the surface fish, it develops during later ontogeny at the adult stage. It is suggested that the eye anlage is developmentally involved in head and brain formation and therefore is still subjected to stabilizing selection. Only after this developmental constraint has been accomplished does variability arise and become obvious (see Sect. 6.21.7) (Figs. 6.43 and 6.48).

### 7.8.2 Loss of Behavioural Traits Not Performed in Darkness

Behavioural traits that depend on vision or light perception provide further strong support for the Neutral Mutation Theory. Such traits are not performed by the surface fish in darkness and their loss does not provide any advantage. Therefore, they are not subjected to selection. Nonetheless, their genetic basis is reduced in the cave fish. This was shown for both the visually triggered aggressive as well as the negative phototactic behaviour (see Sect. 6.12 and 6.14) (Hausberg 1995, Hofmann and Hausberg 1993, Langecker 1989, 2000), the dorsal light reaction (see Sect. 6. 13) (Langecker 1993), and schooling (see Sect. 6.18) (Kowalko et al. 2013a, b; Parzefall 1993).

### 7.8.3 Genetic Studies

In the naked mole rat (*Heterocephalus glaber*), which is no true cave animal but lives underground in the absence of light, out of 200 genes categorized with the GO (gene ontology) term "visual perception", 10% are pseudogenes showing insertion or deletion events (Rétaux and Casane 2013). These include two crystallins (cryBA4 and cryBB3), two out the four vertebrate opsins, and other genes involved in phototransduction and photoreceptor function. In addition, cryGS carries a point mutation. For several of these genes, including the two cited crystallins, a relaxation of functional constraints was noted, as seen through the ratio of non-synonymous substitutions, which surpasses that of synonymous ones. Thus, it seems that genes involved in visual function have been particularly targeted by loss-of-function mutations during the evolution of the naked mole rat genome, suggesting neutral evolution (Fig. 2.6) (Emerling and Springer 2014). A large-scale survey of polymorphism and fixed mutations in the transcriptome of a surface and a cave population of *Astyanax* revealed that a high proportion of the genes carrying mutations responsible for radical amino-acid changes in the cave fish lineage correspond to "eye genes", as deduced from their strong and specific expression in the developing

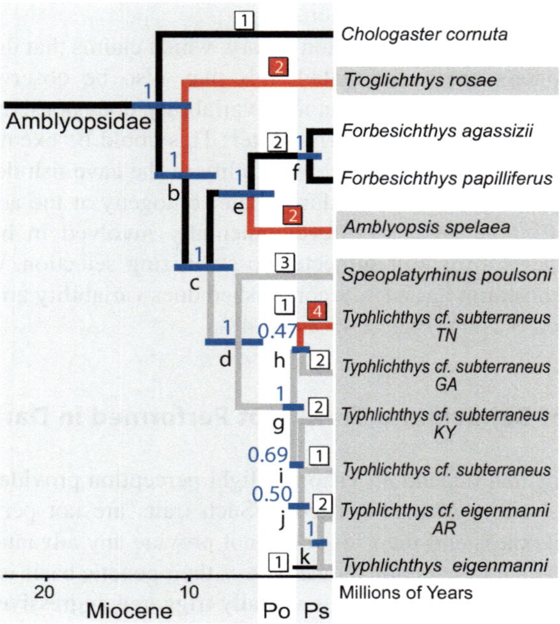


**Fig. 7.3** Chronogram inferred from a multilocus divergence time analysis in amblyopsid cave and surface fish. *Blue bars* at nodes represent 95% highest posterior density intervals of age estimates. Clade posterior probabilities are indicated by *blue numbers* next to nodes. Cave lineages are indicated by *dark gray* tip labels. Branches in *black* are reconstructed as surface and *gray* branches are reconstructed as cave. Nonsynonymous substitutions in *rhodopsin* (*white* or *red square*) are mapped above branches. Lineages with loss-of-function mutations in *rhodopsin* are indicated by *red* branches and *red squares*. *AR* Arkansas, *GA* Georgia, *KY* Kentucky, *Po* to *Ps* Pleistocene to present, *TN* Tennessee (adapted from Niemiller et al. 2012)

visual system in zebrafish (Hinaux et al. 2013). This confirms that eye-related genes are also under relaxed selection in *Astyanax* cavefish.

Further evidence for repeated loss of functional constraint of rhodopsin is also present in amblyopsid cave fishes where, in at least in three cave lineages, loss-of-function mutations have independently accumulated and variability of regressive mutations is extant (Fig. 7.3) (Niemiller et al. 2012). Furthermore, intra- and inter-species analyses suggest that the "blind" clock in *Phreatichthys andruzzii* evolved because of the loss of selective constraints on a trait that was no longer adaptive. Based on this change in selective regimen, it was estimated that the functional constraint on cave fish melanopsin photo pigment (*opn4m2*) was relaxed at ~5.3 million years ago. The visual photoreceptor rhodopsin, expressed in the brain and implicated in the photophobic behaviour of *P. andruzzi*, shows similar evolutionary patterns (Calderoni et al. 2016). Also, in the subterranean bathynellean crustaceans, a reduced opsin repertoire was detected (Kim et al. 2017).

In *Astyanax* cave fish, it was shown that the genetic basis of eye reduction partially diverges in the different cave populations. The same could be shown for the albino and the brown gene (Gross et al. 2009; Protas et al. 2007; Wilkens 1971; Wilkens and Strecker 2003). All this corroborates that most of the regressive mutations originate after cave entry, occur at random, and accumulate after a species has colonized the cave habitat.

Besides such de novo mutations, which characteristically appear after cave colonization, cryptic or standing variation is hypothesized to play a role in the regression of *Astyanax* cave fish traits, also. It is assumed that standing variation already pre-exists in the ancestral surface fish. For example, the melanocortin 4 receptor (*MC4R*) gene contributing to the insatiable appetite of cave fish appears in them due to selection from standing genetic variation already present in surface populations (Aspiras et al. 2015). Also, standing variation for eye size was revealed to exist in *Astyanax* surface fish. Its phenotypic manifestation is usually masked by heat shock protein 90 (HSP90). Blocking HSP90 by the chemical inhibitor radicicol elicits large variation in eye size of larval surface fish. HSP90 is hypothesized to provide a molecular mechanism for buffering genetic variation and release it in response to environmental stress (Rohner et al. 2013).

However, independently from whether the phenotypic variability of regressive traits originates from de novo mutations and/or from standing variation, the alleles are neutral in absolute darkness and can only persist because of the loss of purifying selection. Assuming selection for regressive traits would play a role, variability should not persist for long because selection acts strongly to eliminate it.

## 7.9 Variability and Constructive (Darwin's) Gain

Irrespective of the presence or absence of selection, variability plays the central role in Darwin's concept of evolution. It is generally accepted that this phenomenon is not restricted to regressive processes, as might be the impression now, but is the source of origin and the starting point of constructive evolution too. However, phenotypic variability normally does not arise, because in order to preserve the optimal functional capability of a trait, variability is eliminated by selection.

Nonetheless, examples demonstrating the importance of variability for the evolutionary progress can be found in some rare cases. They are, for example, provided by the intralacustrine radiation of endemic species flocks, as is described for fish like the cichlids in the Great Lakes of East Africa (Barlow 2000; Schön and Martens 2004), the cyprinid genus *Puntius* from the Philippine Lake Lanao, the sculpins from Lake Baikal, the Arctic char (*Salvelinus alpinus*) from Lake Thingvallavatn in Iceland (Jónasson 1992, 1993, 1998), the lake whitefish genus *Coregonus* from lakes in the northern hemisphere (Schluter 1996; Turgeon et al. 1999) or the egg-laying toothcarps (Cyprinodontidae) genus *Orestias* from the Andean Lake Titicaca (Villwock 1986) and the genus *Cyprinodon* from Laguna Chichancanab in the Yucatán in Mexico (Humphries and Miller 1981). Such lakes are either characterized by geographical isolation or extreme abiotic conditions resulting from chemistries or low water temperature leading to depauperate faunas with open niches. Only limited spectrums of one or rarely more fish species are primarily able to colonize such habitats. They encounter an environment that is characterized by the absence of all those species normally occupying the various niches offered by the lake ecosystem. Generally each species is so well adapted to its specific niche that according to Gause's law no other species using the same niche can compete. However, species invading an "empty" ecosystem are no longer restricted to the original niche that they are morphologically and ecologically adapted to. Due to the lack of competing species, interspecific selection is missing and variability may arise. Now disruptive selection can act and generate ecotypic adaptive divergence, for example, in trophic differentiation to exploit different food resources. Such resource-based polymorphism can lead to different morphotypes, which might be the initial step in a speciation process and finally result in different species.

Many examples of intralacustrine speciation in fishes occur in large and geologically old lakes containing complex ecosystems. Due to their age, the process of radiation and speciation is widely advanced or already finished. In contrast, however, the endemic *Cyprinodon* species flock of Laguna Chichancanab, situated in the centre of the Yucatan Peninsula (Mexico), has recently evolved. The lake was probably totally desiccated about 8000 years ago, after which the water level was higher than today for about 5000 years. From 2000 years ago to the present day, lake levels have remained more or less stable (Covich and Stuiver 1974; Hodell et al. 1995, 2001). The Laguna Chichancanab is a small and shallow tropical lake with nearly constant water temperatures throughout the year. It provides an excellent system to study mechanisms of adaptive radiation and speciation because of the simplicity of the lake's ecosystem, its relatively small size, and its young geological age. The water is brackish and saturated with calcium sulphate. Therefore, it is intolerable to most freshwater species (Humphries and Miller 1981; Strecker 2006; Strecker et al. 1996). The invertebrate fauna is low in diversity, and the fish community originally solely comprises an endemic *Cyprinodon* species flock and only one other species, the live-bearing toothcarp *Gambusia sexradiata* (Poeciliidae) (Humphries 1984; Humphries and Miller 1981; Strecker 2006). Mitochondrial DNA data indicate that the species flock is monophyletic and evolutionarily young and has probably evolved after the lake filled up about 8000 years ago. The most likely ancestor, C. *artifrons*, still occurs in coastal lagoons today (Strecker 2006).

The *Cyprinodon* species flock consists of seven species, and a group of specimens amounting to 45–60% of the whole assembly that cannot be assigned unequivocally to one of these species (Strecker 2006). All species co-occur throughout the whole lake. Within the flock, the most common species, C. *beltrani,* is morphologically and ecologically similar to the presumed sister species of the flock, C. *artifrons*. The other members differ strikingly from C. *beltrani,* as well as from each other, mainly by head morphology (Fig. 7.4). These differences suggest trophic divergence and exploitation of different feeding niches by each species. *C. beltrani* is a substrate feeder with a diet consisting

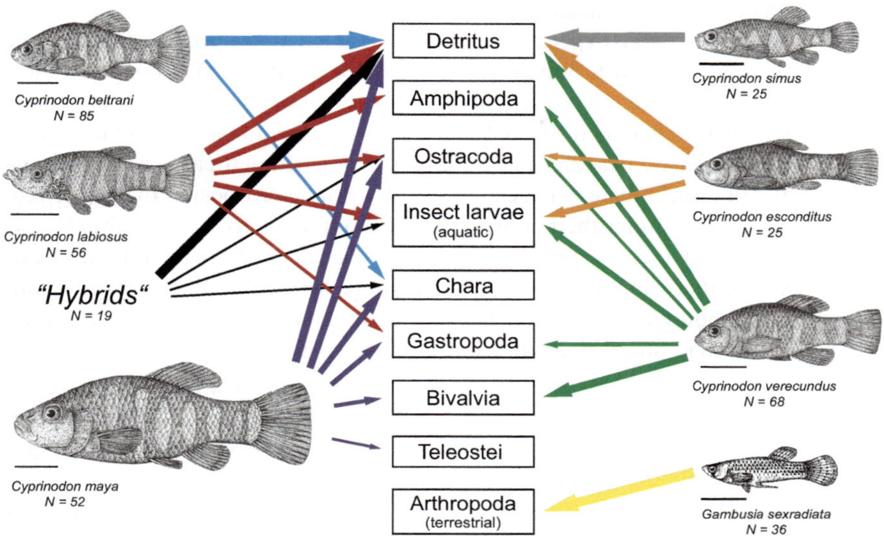


**Fig. 7.4** Prey items of members of the *Cyprinodon* species flock (Cyprinodontidae) and of *Gambusia sexradiata* (Poeciliidae). *Thickness of arrows* corresponds to the mean relative dry weight, values below 1% not shown (adapted from Horstkotte and Strecker 2005)

primarily of detritus. The other six species, C. *maya,* C. *labiosus,* C. *verecundus,* C. *simus,* C. *esconditus,* and C. *suavium,* have a significantly shorter gut indicating a more carnivorous diet (Humphries 1984; Humphries and Miller 1981; Stevenson 1992; Strecker 2002, 2006, Horstkotte and Strecker 2005). Studies of gut content have revealed that all species, like C. *beltrani,* still rely on detritus as an important food item but, in contrast to it, feed to varying degrees on different animal prey. For example, the gut of C. *labiosus* contains a relatively higher amount of amphipods, that of C. *verecundus* of bivalves, and that of C. *maya* of ostracods and gastropods. The group of specimens that cannot be assigned to one of the described species of the flock has a short gut similar to the more carnivorous species (Fig. 7.4), yet, in contrast, there was no apparent preference for any particular food item. Most interestingly, no *Cyprinodon* ecotype has taken advantage of the water surface feeding niche, which was originally occupied by *P. sexradiata* (Horstkotte and Strecker 2005).

Based on mtDNA and microsatellite data, as well as behavioural studies, only the largest species, *C. maya,* shows evidence of reproductive isolation (Strecker 1996, 2006; Strecker et al. 1996; Strecker and Kodric-Brown 2000). For the other endemic species, it is unclear whether they are in the initial phases of speciation and/or if hybridization occurs among them. Female choice experiments examining chemical and visual cues in three of the species, *C. beltrani, C. labiosus*, and *C. maya,* indicate that different levels of reproductive isolation have been reached (Fig. 7.5). These differences are supported by mate choice trials allowing females to spawn with conspecific or heterospecific males (Strecker and Kodric-Brown 2000). From these behavioural experiments it is concluded that species-specific male traits, in combination with female preferences for them, differ among the three species. In *C. maya* there has been sufficient divergence to establish a premating reproductive barrier based on visual and chemical cues alone. The mate choice behaviour between *C. beltrani* and *C. labiosus* is asymmetrical (i.e. *C. beltrani* females do not discriminate between conspecific and heterospecific males whereas *C. labiosus* females prefer their conspecific males). It is likely that the *Cyprinodon* species of Laguna Chichancanab are at different stages of differentiation. The group of *Cyprinodon* that cannot be assigned unequivocally to one of the species may represent morphological plasticity, hybrids, or include as-yet undetected species (Kodric-Brown and Strecker 2001, Strecker et al. 1996).

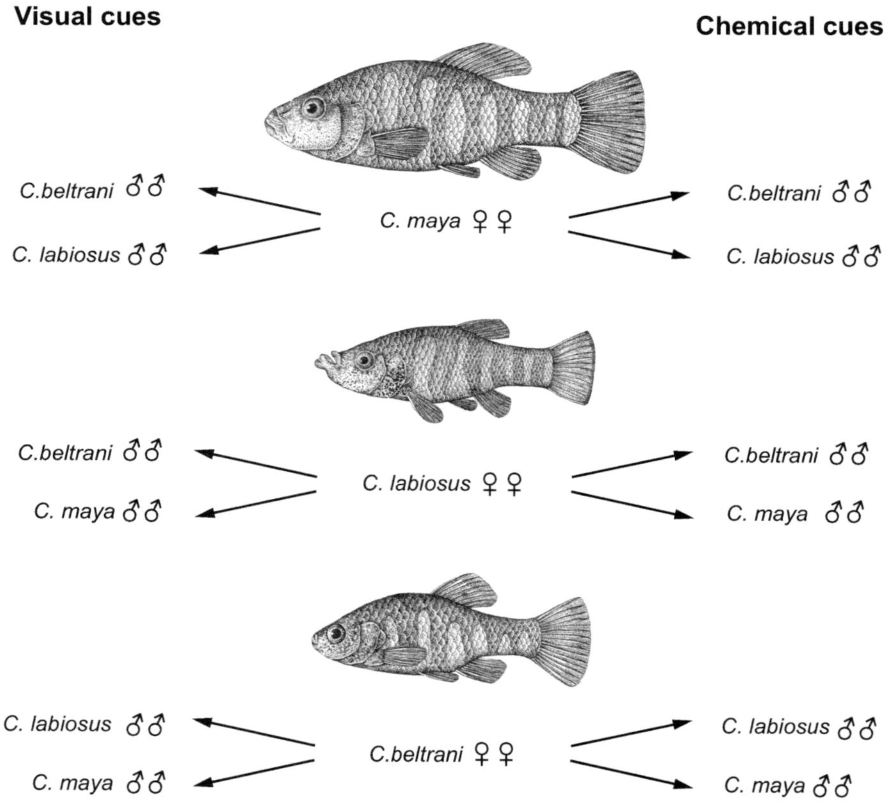


**Fig. 7.5** Summary of responses of *Cyprinodon beltrani, C. labiosus,* and C. *maya* females (Cyprinodontidae) to visual and chemical cues of conspecific and heterospecific males (adapted from Kodric-Brown and Strecker 2001)

## 7.10 Concluding Remarks

The bizarre appearance of the blind and pale phenotype of cave animals might insinuate that the basic principles of rudimentation are exclusive to them. This understanding of a special position is reinforced by the regressive processes in cave animals being interpreted above all as dominated by limited energy supply or by the central role of eye reduction providing a pleiotropic spin-off of constructive evolution of compensatory traits like lateral line and taste senses. Anthropocentrically, cave animals are often even looked upon as suffering from the same pathological phenomena that occur in humans (Pennisi 2016). For example, “sleep loss” (Duboué et al. 2011), “food addiction” (Elipot et al. 2013), “obesity” (Aspiras et al. 2015), or the deleterious risk of the eye being hurt in darkness (Barr 1968) have all been proposed. However, caves in general are not food poor, and nor could the before-mentioned spin offs be proven. Cave animals are no outliers but demonstrate that rudimentation is a universal evolutionary phenomenon, which may in principle concern every trait of all members of each systematic group. For example, eye reduction has also evolved in fossorial mammals, flightless birds have reduced their wings, and the taste for umami got lost when the Giant Panda changed from a carnivorous to a vegetarian diet. Principles found here can also be detected in the human species (Gross and Perkins 2008). Regressive mutants causing eye defects no longer had a selective disadvantage after the transition from hunting to agriculture. To a certain degree, the same happened to human dentition after the development of cooking, barbecuing, and knives. In both traits, the existing high variability indicates relaxed selection. The ambiguous character of some mutational events, that, for example, are found in traits like changed activity patterns in cave animals, was termed adaptive gene loss in the case of the evolution of photosensitivity in vertebrates (Davies 2011). It is also demonstrated by human skin pigmentation in which it was found that light skin colour is the derived state and, like the regressive eye in different *Astyanax* cave populations, is independent in origin in Europeans and Asians, because these mutations occurred at random, whereas dark skin colour seems to reflect the ancestral state (Lao et al. 2007). Regressive mutations of dark pigmentation are neutral in northern latitudes and genetically independently facilitated vitamin D synthesis. In contrast, they were deleterious close to the equator, where they were eliminated by stabilizing selection (Beleza et al. 2012). Cave animals, in particular, when closely related sister surface forms are still available, play a prime role in the study of evolutionary processes.

## References

Aspiras AC, Rohner N, Martineau B et al (2015) Melanocortin 4 receptor mutations contribute to the adaptation of cavefish to nutrient-poor conditions. Proc Natl Acad Sci U S A 112: 9668–9673

Barlow GW (2000) The cichlid fishes. Nature’s grand experiment in evolution. Perseus Publishing, Cambridge

Barr TC (1968) Cave ecology and the evolution of troglobites. In: Dobzhansky T, Hecht MK, Steere WC (eds) Evolutionary biology. Plenum Press, New York, pp 35–101
Barton RA, Harvey PH (2000) Mosaic evolution of brain structure in mammals. Nature 405: 1955–1058
Beleza JC, Santos AM, McEvoy B et al (2012) The timing of pigmentation lightening in Europeans. Mol Biol Evol 30(1):24–35. doi:10.1093/molbev/mss207
Besharse JC, Brandon RA (1974) Effects of continuous light and darkness on the eyes of the troglobitic salamander *Typhlotriton spelaeus*. J Morphol 149:527–546
Bibliowicz J, Alié A, Père S et al (2013) Differences in chemosensory response between eyed and eyeless *Astyanax mexicanus* of the Rio Subterráneo cave. EvoDevo 4:25–31
Bilandžija H, Parkhurst A, Jeffery WR (2013) A potential benefit of albinism in *Astyanax* cavefish: downregulation of the *oca2* gene increases tyrosine and catecholamine levels as an alternative to melanin synthesis. PLoS One 8:1–14
Borowsky RL (2015) Regressive evolution: testing hypotheses of selection and drift. In: Keene AC, Yoshizawa M, McGaugh SE (eds) Biologiy and evolution of the Mexican cavefish. Elsevier, Amsterdam, pp 93–107
Borowsky RL, Cohen D (2013) Genomic consequences of ecological speciation in *Astyanax* cavefish. PLoS One 8(11):e79903. doi:10.1371/journal.pone.0079903
Calderoni L, Rota-Stabelli O, Frigato E et al (2016) Relaxed selective constraints drove functional modifications in peripheral photoreception of the cavefish *P. andruzzii* and provide insight into the time of cave colonization. Heredity 117:383–392
Covich A, Stuiver M (1974) Changes in oxygen 18 as a measure of long-term fluctuations in tropical lake levels and molluscan populations. Limnol Oceanogr 19:682–691
Crish SD, Dengler-Crish C, Catania KC (2006) Central visual system of the naked mole-rat (*Heterocephalus glaber*). Anat Rec A Discov Mol Cell Evol Biol 288(2):205–212
Culver DC (1987) Eye morphometrics of cave and spring populations of *Gammarus minus* (Amphipoda: Gammaridae). J Crustac Biol 7:136–147
Culver DC, Kane TC, Fong DW (1995) Adaptation and natural selection in caves. Harvard University Press, Cambridge
Davies WL (2011) Adaptive gene loss in vertebrates: photosensitivity as a model case. In: Encyclopedia of life sciences (ELS). Wiley, Chichester. doi:10.1002/9780470015902.a0022890
Deharveng L, Bedos A (2000) The cave fauna of Southeast Asia. Origin, evolution and ecology. In: Wilkens H, Culver DC, Humphreys WF (eds) Ecosystems of the world: subterranean ecosystems, vol 30. Elsevier, Amsterdam, pp 603–632
Dillman CB, Bergstrom DE, Noltie DB et al (2010) Regressive progression, progressive regression or neither? Phylogeny and evolution of the Percopsiformes (Teleostei, Paracanthopterygii). Zool Scr 40:45–60
Dollo L (1893) Les lois de l'évolution. Bull Soc Belge Geol Pal Hydr VII:164–166
Duboué ER, Keene AC, Borowsky RL (2011) Evolutionary convergence on sleep loss in cavefish populations. Curr Biol 21:671–676
Eigenmann CH (1909) Cave vertebrates of America. Carnegie Inst Wash Publ 104:1–241
Elipot Y, Hinaux H, Callebert J et al (2013) Evolutionary shift from fighting to foraging in blind cavefish through changes in the serotonin network. Curr Biol 23:1–10
Ellis R (1996) Deep Atlantic: life, death, and exploration in the abyss. The Lions Press, New York
Emerling CA, Springer MS (2014) Eyes underground: regression of visual protein networks in subterranean mammals. Mol Phylogenet Evol 78:260–270
Espinasa L, Borowsky RL (2000) Eyed cave fish in a karst window. J Cave Karst Stud 62:180–183
Fernandes CS, Batalha MA, Bichuette ME (2016) Does the cave environment reduce functional diversity? PLoS One 11(3):e0151958. doi:10.1371/journal.pone.0151958
Freese E, Yoshida A (1965) The role of mutations in evolution. In: Bryson V, Vogel HJ (eds) Evolving genes and proteins. Academic, New York, pp 341–355

Gnaspini P, Trajano E (2000) Guano communities in tropical caves. In: Wilkens H, Culver DC, Humphreys WF (eds) Ecosystems of the world: subterranean ecosystems, vol 30. Elsevier, Amsterdam, pp 251–268

Gross JM, Perkins BD (2008) Zebrafish mutants as models for congenital ocular disorders in humans. Mol Reprod Dev 75:547–555

Gross JB, Borowsky RL, Tabin CJ (2009) A novel role for *Mc1r* in the parallel evolution of depigmentation in independent populations of the cave fish, *Astyanax mexicanus*. PLoS Genet 5:1–14

Hausberg C (1995) Das Aggressionsverhalten von *Astyanax fasciatus* (Cuvier 1819; Characidae, Teleostei): Ontogenie, Genetik und Evolution bei der epigäischen und hypogäischen Form. Dissertation, University of Hamburg

Hinaux H, Poulain J, Da Silva C et al (2013) De novo sequencing of *Astyanax mexicanus* surface fish and Pachón cavefish transcriptomes reveals enrichment of mutations in cavefish putative eye genes. PLoS One 8(1):e53553. doi:10.1371/journal.pone.0053553

Hinaux H, Blin M, Fumey J et al (2015) Lens defects in Astyanax mexicanus cavefish: evolution of crystallins and a role for alphaA-crystallin. Dev Neurobiol 75(5):505–521

Hinaux H, Devos L, Blin M et al (2016) Sensory evolution in blind cavefish is driven by early embryonic events during gastrulation and neurulation. Development 143:4521–4532. doi:10.1242/dev.141291

Hoch H (2000) Acoustic communication in darkness. In: Wilkens H, Culver DC, Humphreys WF (eds) Ecosystems of the world: subterranean ecosystems, vol 30. Elsevier, Amsterdam, pp 211–220

Hodell DA, Curtis JH, Brenner M (1995) Possible role of climate in the collapse of classic maya civilization. Nature 375:391–394

Hodell DA, Brenner M, Curtis JH et al (2001) Solar forcing of drought frequency in the Maya lowlands. Science 292:1367–1370

Hoffman S, Hausberg C (1993) The aggressive behavior of the Micos cave population *(Astyanax fasciatus,* Characidae, Teleostei) after selection for functional eyes in comparison to an epigean one. Mémoires de Biospéléologie 20:101–103

Horstkotte J, Strecker U (2005) Trophic differentiation in the phylogenetically young *Cyprinodon* species flock (Cyprinodontidae, Teleostei) from Laguna Chichancanab (Mexico). Biol J Linn Soc 85:125–134

Huber R, Staaden MJ van, Kaufman LS et al (1997) Microhabitat use, trophic patterns, and the evolution of brain structure in African cichlids. Brain Behav Evol 50:167–182

Humphries JM (1984) *Cyprinodon verecundus* n. sp., a fifth species of pupfish from Laguna Chichancanab. Copeia 1984:58–68

Humphries JM, Miller RR (1981) A remarkable species flock of *Cyprinodon* from Lake Chichancanab, Yukatan, Mexico. Copeia 1981:52–64

Hüppop K (1989) Genetic analysis of oxygen consumption rate in cave and surface fish of *Astyanax fasciatus* (Characidae, Pisces). Further support for the neutral mutation theory. Mémoires de Biospéléologie 16:163–168

Hüppop K (2000) How do cave animals cope with the food scarcity in caves? In: Wilkens H, Culver DC, Humphreys WF (eds) Ecosystems of the world: subterranean ecosystems, vol 30. Elsevier, Amsterdam, pp 159–188

Hüppop K (2012) Adaptation to low food. In: White WB, Culver DC (eds) Encyclopedia of caves, 2nd edn. Elsevier, Amsterdam, pp 1–9

Hüppop K, Wilkens H (1991) Bigger eggs in subterranean *Astyanax fasciatus*. Z Zool Syst Evol 29:280–288

Jeffery WR (2005) Adaptive evolution of eye degeneration in the Mexican blind cavefish. J Hered 96:185–196

Jeffery WR, Ma L, Parkhurst A et al (2015) Pigment regression and albinism in *Astyanax* cavefish. In: Keene AC, Yoshizawa M, McGaugh SE (eds) Biology and evolution of the Mexican cavefish. Elsevier, Amsterdam, pp 155–171

Jónasson PM (1992) The ecosystem of Thingvallavatn: a synthesis. Oikos 64:405–434
Jónasson PM (1993) Continental rifting-energy pathways of two contrasting rift lakes. Verhandlungen des Internationalen Vereins für Limnologie 25:1–14
Jónasson PM (1998) Continental rifting and habitat formation: Arena for resource polymorphism in Arctic charr. Ambio 27:162–169
Juberthie-Jupeau L (1976) Sur le système neurosécréteur du pédoncule oculaire d'un décapode souterrain microphthalme *Typhlatya garciai* Chace. Ann Spéléol 31:107–114
Katz PS (2011) Neural mechanisms underlying the evolvability of behavior. Philos Trans R Soc Lond B Biol Sci 366:2086–2090
Katz PS, Lillvis JL (2014) Reconciling the deep homology of neuromodulation with the evolution of behavior. Curr Opin Neurobiol 29:39–47. doi:10.1016/j.conb.2014.05.002
Kim B-M, Kang S, Ahn DH et al (2017) First insights into the subterranean crustacean Bathynellacea transcriptome: transcriptionally reduced opsin repertoire and evidence of conserved homeostasis regulatory mechanisms. PLoS One 12(1):e0170424. doi:10.1371/journal
Kimura M (1968) Evolutionary rate at the molecular level. Nature 217:624–626
King JL, Jukes TH (1969) Non-Darwinian evolution. Science 164:788–797
Kodric-Brown A, Strecker U (2001) Responses of *Cyprinodon maya* and *C. labiosus* females to visual and olfactory cues of conspecific and heterospecific males. Biol J Linn Soc 74:541–548
Konec M, Delić T, Trontelj P (2016) DNA barcoding sheds light on hidden subterranean boundary between Adriatic and Danubian drainage basins. Ecohydrology 9:1304–1312. doi:10.1002/eco.1727
Kosswig C (1935) Die Evolution von "Anpassungs"-Merkmalen bei Höhlentieren in genetischer Betrachtung. Zool Anz 112:148–155
Kosswig C (1949) Phänomene der regressiven Evolution im Lichte der Genetik. Communications de la Faculté des Sciences de l'Université d'Ankara 8:110–150
Kosswig C (1960a) Darwin und die degenerative evolution. Abhandlungen und Verhandlungen des Naturwissenschaftlichen Vereins Hamburg NF 4:21–42
Kosswig C (1960b) Zur Phylogenese sogenannter Anpassungsmerkmale bei Höhlentieren. Int Rev Ges Hydrobiol 45:493–512
Kosswig C (1976) Génétique et l'évolution regressive - Mécanismes de la rudimentation des organes chez les embryons de vertébrés. Colloq Int CNRSN 266:19–29
Kosswig C, Kosswig L (1940) Die Variabilität bei *Asellus aquaticus*, unter besonderer Berücksichtigung der Variabilität in isolierten unter- und oberirdischen Populationen. Révue de la Faculté des Sciences Istanbul 5:1–55
Kowalko JE, Rohner N, Linden TA et al (2013a) Convergence in feeding posture occurs through different loci in independently evolved cave populations of *Astyanax mexicanus*. Proc Natl Acad Sci U S A 110:16933–16938
Kowalko JE, Rohner N, Rompani SB et al (2013b) Loss of schooling behavior in cavefish through sight-independent and sight-dependent mechanisms. Curr Biol 23:1874–1883
Krishnan J, Rohner N (2017) Cavefish and the basis for eye loss. Philos Trans R Soc B 372(1713):20150487. doi:10.1098/rstb.2015.0487
Langecker TG (1989) Studies on the light reaction of epigean and cave populations of *Astyanax fasciatus* (Characidae, Pisces). Mem Biospeleol 16:169–176
Langecker TG (1993) Genetic analysis of the dorsal light reaction in epigean and cave-dwelling *Astynnax fasciatus* (Teleostei, Characidae). Ethol Ecol Evol 3:357–364
Langecker TG (2000) The effects of continuous darkness on cave ecology and cavernicolous evolution. In: Wilkens H, Culver DC, Humphreys WF (eds) Ecosystems of the world: subterranean ecosystems, vol 30. Elsevier, Amsterdam, pp 135–157
Langecker TG, Neumann B, Hausberg C et al (1995) Evolution of the optical releasers for aggressive behavior in cave-dwelling *Astyanax fasciatus* (Teleostei, Characidae). Behav Process 34:161–168

Langecker TG, Wilkens H, Parzefall J (1996) Studies of the trophic structure of an energy-rich Mexican cave (Cueva de las Sardinas) containing sulfurous water. Mém Biospéléol 23: 121–125

Lao O, de Gruijter JM, Duijn K v et al (2007) Signatures of positive selection in genes associated with human skin pigmentation as revealed from analyses of single nucleotide polymorphisms. Ann Hum Genet 71:354–369

Menuet A, Alunni A, Joly JS et al (2007) Expanded expression of Sonic Hedgehog in *Astyanax* cavefish: multiple consequences on forebrain development and evolution. Development 134:845–855. doi:10.1242/dev.02780

Meyer-Rochow VB, Juberthie-Jupeau L (1987) An electron microscope study of the eye of the cave mysid Heteromysoides cotti from the island of Lanzarote (Canary Islands). Stygologia 3:24–33

Mitchell RW, Russell WH, Elliott WR (1977) Mexican eyeless characin fishes, genus *Astyanax*: environment, distribution, and evolution. Special publications of the Museum Texas Tech University 12:1–89

Moran D, Softley R, Warrant EJ (2015) The energetic cost of vision and the evolution of eyeless Mexican cavefish. Sci Adv 1(8):e1500363. doi:10.1126/sciadv.1500363

Muller HJ (1949) The Darwinian and modern conceptions of natural selection. Proc Am Philos Soc 93:459–470

Nei M (2013) Mutation-driven evolution. Oxford University Press, Oxford

Nei M, Suzuki Y, Nozawa M (2010) The neutral theory of molecular evolution in the genomic era. Annu Rev Genomics Hum Genet 11:265–289

Niemiller ML, Fitzpatrick BM, Shah P et al (2012) Evidence for repeated loss of selective constraint in rhodopsin of amblyopsid cavefishes (Teleostei: Amblyopsidae). Evolution 67:732–748

Niven JE, Anderson JC, Laughlin SB (2007) Fly photoreceptors demonstrate energy-information trade-offs in neural coding. PLoS Biol 5:828–840

Orr HA (1998) Testing natural selection vs. genetic drift in phenotypic evolution using quantitative trait loci. Genetics 149:2099–2104

Palacios M, Voelker G, Rodriguez LA et al (2016) Phylogenetic analyses of the subgenus *Mollienesia* (Poecilia, Poeciliidae, Teleostei) reveal taxonomic inconsistencies, cryptic biodiversity, and spatio-temporal aspects of diversification in Middle America. Mol Phylogenet Evol 103:230–244

Parzefall J (1993) Schooling behaviour in population-hybrids of *Astyanax fasciatus* and *Poecilia mexicana* (Pisces, Characidae and Poeciliidae). In: Schröder JH, Bauer J (eds) Trends in ichthyology: an international perspective. Blackwell Scientific, Oxford, pp 297–303

Parzefall J (2001) A review of morphological and behavioural changes in the cave molly, *Poecilia mexicana*, from Tabasco, Mexico. Environ Biol Fishes 62:263–275

Pennisi E (2016) Blind cave fish may provide insights into human health. Science 352(6293): 1502–1503. doi:10.1126/science.352.6293.1502

Peters N, Peters G (1968) Zur genetischen interpretation morphologischer Gesetzmäßigkeiten der degenerativen evolution. Zeitschrift Morphologie der Tiere 62:211–244

Peters N, Peters G (1973) Genetic problems in the regressive evolution of cavernicolous fish. In: Schröder JH (ed) Genetics and mutagenesis of fish. Springer, Berlin, pp 187–201

Peters N, Scholl A, Wilkens H (1975) Der Micos-Fisch, Höhlenfisch in statu nascendi oder Bastard? Ein Beitrag zur Evolution der Höhlentiere. J Zool Syst Evol Res 13:110–124

Peters N, Schacht V, Schmidt W et al (1993) Gehirnproportionen und Ausprägungsgrad der Sinnesorgane von *Astyanax mexicanus* (Pisces, Characinidae). Z Zool Syst Evol 31:144–159

Poulson TL, White WB (1969) The cave environment: limestone caves provide unique natural laboratc for studying biological and geological processes. Science 165:971–981

Protas M, Conrad M, Gross JB et al (2007) Regressive evolution in the Mexican Cave Tetra, *Astyanax mexicanus*. Curr Biol 17:452–454. doi:10.1016/j.cub.2007.01.051

Protas ME, Trontelj P, Patela NH (2011) Genetic basis of eye and pigment loss in the cave crustacean, *Asellus aquaticus*. Proc Natl Acad Sci U S A 108:5702–5707. doi:10.1073/pnas.1013850108

Rétaux S, Casane D (2013) Evolution of eye development in the darkness of caves: adaptation, drift, or both? EvoDevo 4(1):1–12

Rétaux S, Pottin K, Alunni A (2008) *Shh* and forebrain evolution in the blind cavefish *Astyanax mexicanus*. Biol Cell 100:139–147

Rodrigues FR (2013) *Comparison of brain and cranial nerve morphology between eyed surface fish and blind cave fish of the species Astyanax mexicanus*. Dissertation, Universidade de Lisboa

Rohner N, Jarosz DF, Kowalko JE et al (2013) Cryptic variation in morphological evolution: HSP90 as a capacitor for loss of eyes in cavefish. Science 342:1372–1375

Schemmel C (1967) Vergleichende Untersuchungen an den Hautsinnesorganen ober- und unterirdisch lebender *Astyanax*-Formen. Zeitschrift Morphologie der Tiere 61:255–316

Schemmel C (1980) Studies on the genetics of feeding behavior in the cave fish *Astyanax mexicanus* f. Anopthichthys. An example of apparent monofactorial inheritance by polygenes. Z Tierpsychol 53:9–22

Schluter D (1996) Ecological speciation in postglacial fishes. Philos Trans Biol Sci 351:807–814

Schön I, Martens K (2004) Adaptive, pre-adaptive and non-adaptive components of radiations in ancient lakes: a review. Organ Divers Evol 4:137–156

Stevenson MM (1992) Food habits within the Laguna Chichancanab *Cyprinodon* (Pisces: Cyprinodontidae) species flock. Southwest Nat 37:337–343

Strecker U (1996) Molekulargenetische und ethologische Untersuchungen zur Speziation eines Artenschwarmes der Gattung *Cyprinodon* (Cyprinodontidae, Teleostei). Universität, Hamburg

Strecker U (2002) *Cyprinodon esconditus,* a new pupfish from Laguna Chichancanab, Yucatan, Mexico (Cyprinodontidae). Cybium 26:301–307

Strecker U (2006) Genetic differentiation and reproductive isolation in a *Cyprinodon* fish species flock from Laguna Chichancanab, Mexico. Mol Phylogenet Evol 39:865–872

Strecker U, Kodric-Brown A (2000) Mating preferences in a species flock of Mexican pupfishes (Cyprinodon, Teleostei). Biol J Linn Soc 71:677–687

Strecker U, Meyer CG, Sturmbauer C et al (1996) Genetic divergence and speciation in an extremely young species flock in Mexico formed by the genus *Cyprinodon* (Cyprinodontidae, Teleostei). Mol Phylogenet Evol 6:143–149

Strecker U, Hausdorf B, Wilkens H (2012) Parallel speciation in *Astyanax* cave fish (Teleostei) in Northern Mexico. Mol Phylogenet Evol 62:62–70

Strickler AG, Yamamoto Y, Jeffery WB (2007) The lens controls cell survival in the retina: evidence from the blind cavefish *Astyanax*. Dev Biol 311:512–523

Turgeon J, Estoup A, Bernatchez L (1999) Species flock in the North American Great Lakes: molecular ecology of Lake Nipigon ciscoes (Teleostei, Coregonidae: *Coregonus*). Evolution 53:1857–1871

Verovnik R, Prevorčnik S, Jugovic J (2009) Description of a neotype for *Asellus aquaticus* Linné, 1758 (Crustacea: Isopoda: Asellidae), with description of a new subterranean Asellus species from Europe. Zool Anz - J Comp Zool 248(2):101–118

Villwock W (1986) Speciation and adaptive radiation in Andean *Orestias* fishes. In: Vuilleumier F, Monasterio MM (eds) High altitude tropical biogeography. University Press, Oxford, pp 387–403

Wilkens H (1971) Genetic interpretation of regressive evolutionary processes: studies on hybrid eyes of two *Astyanax* cave populations (Characidae, Pisces). Evolution 25:530–544

Wilkens H (1976) Genotypic and phenotypic variability in cave animals. Studies on a phylogenetically young cave population of *Astyanax mexicanus* (Fillippi). Annales de Spéléologie 3:137–148

Wilkens H (1988) Evolution and genetics of epigean and cave *Astyanax fasciatus* (Characidae, Pisces). Support for the Neutral Mutation Theory. In: Hecht MK, Wallace B (eds) Evolutionary biology, vol 23. Plenum, New York, pp 271–367

Wilkens H (2001) Convergent adaptations to cave life in the *Rhamdia laticauda* catfish group (Pimelodidae, Teleostei). Environ Biol Fishes 62:251–261

Wilkens H (2010) Genes, modules and the evolution of cave fish. Heredity 105:413–422

Wilkens H (2016) Genetics and hybridization in surface and cave *Astyanax* (Teleostei): a comparison of regressive and constructive traits. Biol J Linn Soc 118:911–928

Wilkens H, Strecker U (2003) Convergent evolution of the cave fish *Astyanax* (Characidae, Teleostei): genetic evidence from reduced eye size and pigmentation. Biol J Linn Soc 80: 545–554

Wilkens H, Peters N, Schemmel C (1979) Gesetzmäßigkeiten der regressiven Evolution. Verh Dtsch Zool Ges 1979:123–140

Wilkens H, Strecker U, Yager J (1989) Eye reduction and phylogenetic age in ophidiiform cave fish. Z Zool Syst Evol 27:126–134

Yamamoto Y, Stock DW, Jeffery WR (2004) Hedgehog signalling controls eye degeneration in blind cavefish. Nature 431:844–847

Yamamoto Y, Byerly MS, Jackman WR et al (2009) Pleiotropic functions of embryonic sonic hedgehog expression link jaw and taste bud amplification with eye loss during cavefish evolution. Dev Biol 330:200–211

Yang X-L, Tornquist K, Dowling JE (1988a) Modulation of cone horizontal cell activity in the teleost fish retina. I. Effects of prolonged darkness and background illumination on light responsiveness. J Neurosci 8:2259–2268

Yang X-L, Tornquist K, Dowling JE (1988b) Modulation of cone horizontal cell activity in the teleost fish retina. II. Role of interplexifoem cells and dopamine in regulating light responsiveness. J Neurosci 8:2269–2278

Yoshizawa M, Yamamoto Y, O'Quin KE et al (2012) Evolution of an adaptive behavior and its sensory receptors promotes eye regression in blind cavefish. BMC Biol 10:108–123

Yoshizawa M, Robinson BG, Duboué ER (2015) Distinct genetic architecture underlies the emergence of sleep loss and prey-seeking behavior in the Mexican cavefish. BMC Biol 13: 15–27

Zeutzius I, Probst W, Rahmann H (1984) Influence of dark-rearing on the ontogenetic development of Sarotherodon mossambicus (Cichlidae, Teleostei): II. Effects on allometric growth relations and differentiation of the optic tectum. Exp Biol 43:87–96

GPSR Compliance

*The European Union's (EU) General Product Safety Regulation (GPSR) is a set of rules that requires consumer products to be safe and our obligations to ensure this.*

*If you have any concerns about our products, you can contact us on ProductSafety@springernature.com*

In case Publisher is established outside the EU, the EU authorized representative is:

Springer Nature Customer Service Center GmbH
Europaplatz 3
69115 Heidelberg, Germany

**Batch number: 10755216**

Printed by Printforce, the Netherlands